A CALL TO FARMS

A
CALL
TO
FARMS

Reconnecting to Nature, Food, and
Community in a Modern World

Jennifer Grayson

Countryman Press

An Imprint of W. W. Norton & Company
Independent Publishers Since 1923

For information about permission to reproduce selections from this book, write to Permissions, Countryman Press, 500 Fifth Avenue, New York, NY 10110

For information about special discounts for bulk purchases, please contact W. W. Norton Special Sales at specialsales@wwnorton.com or 800-233-4830

Manufacturing by Lakeside Book Company
Book design by Beth Steidle
Production manager: Julia Druskin

Countryman Press
www.countrymanpress.com

An imprint of W. W. Norton & Company, Inc.
500 Fifth Avenue, New York, NY 10110
www.wwnorton.com

978-1-68268-846-5

10 9 8 7 6 5 4 3 2 1

For Matthew,
whose love has regenerated me

CONTENTS

INTRODUCTION

ADMITTEDLY, THIS BOOK BEGAN as my apocalyptic survival fantasy. Instead, like any good mythic quest, it morphed into a story of hope—and a narrative far greater than my own.

Two years before the dawn of the COVID-19 pandemic, I was researching a different book idea and hosting a podcast, *Uncivilize*, about rewilding—a subculture of the better-known land conservation movement where people pursue a preindustrial or even preagricultural, hunter-gatherer existence. My interviews included survivalists living on a tropical island, primitive skills enthusiasts creating forest schools, and subsistence homesteaders. I've lived in cities my entire adult life, so it doesn't take a psychologist to unpack my personal attraction to the idea of backpedaling from the increasing overwhelm of life in the twenty-first century: the incessant infiltration of technology and media; social isolation and loneliness; disconnection from nature, especially its troubling impact on our kids; escalating global conflict; and accelerating natural disasters validating our fears that the endgame of climate change is not only inevitable, but happening now. In fact, just in the months since I turned in this manuscript, Canada's worst wildfire season on record choked North America in smoke; biblical floods in the Northeast washed away roads and bridges; and unprecedented drought on a landscape damaged by extractive agriculture primed a fire that razed a Maui town to ashes. As I write this, a hurricane is on its trajectory to my city of Los Angeles, where we used to simply fret about traffic and earthquakes.

None of these climatic (and climactic) events, however, would have come as a surprise to my pre-pandemic self, engrossed in research about preppers and the demise of bygone civilizations—although I imagined our own society unraveling in a nebulous, distant future. Still, as time went on, I became a little weary of the doomsday preoccupation. More important, I was unsure of its helpfulness. My frequent musings on the topic would reliably elicit a Debbie Downer trombone-like "wah-wah" sound effect from family and friends. Everyone can feel the tumult of these times but very few of us, myself included, have the wherewithal or the chutzpah to toss aside everything they've ever known and hunt and forage from a cabin in the woods. Some of the solutions being touted in the world of rewilding were inspiring—and even made it into this book—but I wished for a doable purpose in the here and now; preferably one where I would feel more alive and useful than I did rhapsodizing in front of a computer.

I also had a concurrent realization: In my longing to reclaim the ways of the past, it was traditional food culture that most lit my fire. (I came to this revelation one day making a pit stop for *sheng jian bao* in Los Angeles's San Gabriel Valley, a Chinese American enclave renowned for its cuisine. That first crispy, juicy bite outshone the entire day I had just spent at a survival skills course in the nearby mountains.)

And so on January 2, 2020, I aired my last podcast episode and relegated myself to fantasizing how I, like my interviewees, might one day learn how to grow all my own food while ensconced in nature. Shortly after, I shelved the accompanying book project about readying for civilization's collapse because, well, the imminent pandemic apocalypse was kind of mucking up my plans for on-the-ground reporting.

I don't want to spoil too much of the story ahead, but the sound-bite version is that six months into COVID lockdown in Los Angeles, my husband and I decided, *Enough with the daydreaming*, and essentially sold everything we owned and moved with our two young daughters to Central Oregon, where I serendipitously stumbled into the area's local food movement and subsequently enrolled in a groundbreaking farmer training program. The immersive internship was centered around regenerative agriculture—a new (but actually, ancestral) and holistic approach to growing food that restores soil and biodiversity and sequesters carbon in the ground.

I've covered the ills of our industrialized food system for more than a decade, so regenerative farming was a field I was closely following. High-profile books and documentaries were pointing to its promise while sounding the alarm on the finiteness of intensive agriculture—warning of vanishing groundwater and the world's dwindling supply of usable topsoil. Yet until I encountered the training program in Oregon (which you'll read about in the first chapter), it never occurred to me to actually take matters into my own hands and consider small, sustainable farming as a viable career path.

A week into my first farm job, I realized it was the most joyful and fulfilling work I had ever experienced. After two months of being outside all day, nearly every day, I felt the best—both physically and mentally—that I ever had in my life. But the real transformation occurred as I began to meet and learn about the new and driven farmers, graziers, and food activists emerging all over the country. They hadn't grown up in farming families; they came from backgrounds vastly underrepresented in agriculture; and many of them were far younger than I was, not to mention decades younger than the average American farmer. I was awestruck by their intention and ingenuity. They hadn't turned to this way of life as some back-to-the-land fantasy. They had chosen sustainable agriculture as a tactile way to effect environmental activism and food justice; for cultural reclamation; to reconnect to nature, food, and community; to live aligned with their values; to do, in the words of one farmer you'll meet in this book, "something that *means* something."

And during the environmental and societal reckoning of the pandemic—not to mention the collapse of the industrial food supply chain—the work of these regenerative farmers became more meaningful than ever before. They filled the void amid empty supermarket shelves and miles-long food lines, and fed millions of Americans not just food, but the most delicious food many of us had ever tasted. They witnessed hundreds of thousands of people needlessly dying of COVID due to diet-related disparities and pushed ahead for funding and food sovereignty. So I started to wonder: How could we scale a "greatest generation" of sustainable small farmers? What would this country look like transformed by a vast network of resilient local food systems that restore the environment and ensure healthy, fresh food is accessible to all?

These two questions launched me on the journey to write this book. But it was only later that I learned of their urgency.

In the coming decade, 400 million acres of American farmland—*nearly half of all farmland in the United States*—will become available as the older generation of American farmers retires or dies. Meanwhile, the groundswell of new growers eager to steward that land are up against seemingly every obstacle: access to affordable land, access to capital, a livable income, and the billionaires and corporations now grabbing farmland at a staggering pace.

I'm grateful for the fantastical romanticism that buoyed me into the world of farming, otherwise I might not have written this book.

While I now know my driving questions were those only a newbie farmer would ask, I found they aren't a pipe dream: Big Ag may be the norm in the United States, but small growers globally produce around a third of the world's food on farms of five acres or less. Mapping research shows up to 90 percent of Americans could be fed entirely with food raised within 100 miles of where they live. Project Regeneration highlights regenerative agriculture and other nature-based farming methods as critical strategies in the plan to *reverse global warming*. And the human power exists: The number of new, beginning, and young farmers has been increasing for the past ten years, a trend unparalleled in the last century. At the same time, we are in the Great Resignation, where millions of disaffected Americans are leaving their status quo jobs in search of more fulfilling work.

I came to farming as an outsider, and that's exactly the point. Two hundred years ago, nearly all of us lived and worked on the land that fed us.[1] Even a hundred years ago, one-third of us did. Today, that number stands at 1 percent. Yet right now so many of us are yearning for something we can't name, an intangible we don't even realize has been lost. It's our connection to our food, that most fundamental of human needs, and it is that which ties us to everything else. In the words of Anthony Bourdain, "I think food, culture, people, and landscape are all absolutely inseparable."

Maybe you feel the stirring, too, and that's what led you to these pages. This is the book I wish someone had handed me ten years ago, when I was searching for a meaningful way forward. These are the stories of a new, diverse generation of agrarians unfolding an alternate vision of the future, if only more of us would join the call.

---

1 Not all of our own free will.

ONE

Mahonia Gardens

"After the bare requisites to living and reproducing, man wants most to leave some record of himself, a proof, perhaps, that he has really existed. He leaves his proof on wood, on stone, or on the lives of other people."

—John Steinbeck, *The Pastures of Heaven*

THERE WOULD BE DAYS ON THE FARM WHERE I FELT I WAS HELD IN the palm of the earth. I'd kneel on the soil, bundled in my fleece, and pluck perfect white turnips from the land. I'd watch ponderosa pines swaying in the frame of puffy clouds and a snowcapped mountain and think, *This is what it means to live.*

But this would not be one of those days.

It was the first day of my summer farming internship at Mahonia Gardens, a one-acre vegetable farm tucked into the high desert wilderness of Sisters, Oregon. It was thirty-seven degrees, and a hailstorm had blasted across the mountains the night before. I was living in the nearby city of Bend and had risen before dawn to drive there through sheets of rain on a twenty-two-mile snake of highway, squeezing the steering wheel each time a semi roared past. From the time I parked at the farm's fence and made my way through the downpour to the open-air pack shed, my jeans were clinging to my legs like a wet paper towel.

Benji Nagel, who started Mahonia Gardens in 2013 with his now wife, Carys (sounds like "carrots") Wilkins, welcomed me cheerfully by the washtub, then regarded me with wordless consideration. He was wearing a full rainsuit. I cinched the hood of my windbreaker and clocked my rookie mistake: *There's no bad weather on the farm, only bad clothing.* I followed his lean frame out to the field for my first farming lesson, harvesting salad greens that I had, until that point, been buying neatly bagged at my local market. The deluge diminished to a drizzle. We crouched in the mud alongside glistening beds of red and green leaf lettuce, mizuna, and Koji tatsoi, and Benji showed me how to grasp each head like a ponytail and, with a satisfying slice from a tiny plastic-handled knife, lop the leaves above the root until we had twenty-five pounds' worth in giant blue containers.

Back at the shed, Benji demonstrated how to wash the salad: swishing it in porcelain basins of icy water, fishing the leaves out with a trout net, then loading them into two old-school Maytag washing machines to dry on the spin cycle (a common hack among small farmers). Then, as Benji reached into the drum to retrieve the salad mix, a blast of wind tore through the shed. He stood up hastily, arms full of greens, just in time to see his rain pants split up the crotch. So, on that first day, I learned three pivotal lessons of small-scale farming in one of the harshest climates in the lower forty-eight states: dress well, think outside the box, and have a sense of humor.

I FIRST HEARD OF Carys and Benji when I saw their photo on a "Meet Our Farmers" wall at the nonprofit market Central Oregon Locavore. Whereas the average age of the American farmer is fifty-eight years old, they were young—then, late twenties—and their faces were beaming. I later learned they had a three-year-old son, Junius, named for the character Junius Maltby, who lives in bucolic bliss in John Steinbeck's *The Pastures of Heaven*. The summer of my internship, in 2021, Carys, thirty-three, would be pregnant with their second child.

Staring at that Meet Our Farmers wall was the first time I saw the term *market garden*, and I was astonished to read that Carys and Benji grew such abundance—some forty varieties of vegetables, herbs, and flowers—on a single acre, with no pesticides or artificial fertilizers and only a smattering of hand tools. My farming experience until that point consisted of helming a lone raised bed in a sea of pavement at my daughters' Los Angeles public schools, and it had been a while since my city brain could envision the size of an acre. And so, when I first met with Benji and Carys on a twenty-two-degree December morning, socially distanced outside a coffee shop to hear about farm life, I envisioned Benji with his red-blond whiskers and Carys in her cherry-colored cap, huddled over cedar beds in a backyard operation, the two delicately tending veggies with hand tools the size of dinner forks.

My first day at Mahonia Gardens, I nearly laughed out loud at how ridiculously I underestimated market gardening, a biointensive method of farming that produces high yields of crops on minimal land. The farm's open acre unfurled before me with in-ground bed after bed of lettuces, chard, kale, spinach, turnips, radishes, beets, kohlrabi, onions, carrots, and potatoes. It was an ever-rotating bounty over the course of the season, according to plans plotted out months in advance in a notebook that Carys and Benji called "The Brain." Each bed was spaced just far enough apart from the next for me to accidentally clomp backward on the bed of onions while learning to harvest carrots in my first week. (Rule number one of market gardening: Don't step on the beds. Compacted soil harms plant growth.) I also promptly realized that "hand tools" simply meant ones without a motor and were considerably larger than dinner fork–sized. After a month of blasting through weeds with a push contraption called a wheel hoe, my arms were as ripped as an Ashtanga instructor's.

While market gardening has been recently popularized by Quebecois farmer Jean-Martin Fortier, author of *The Market Gardener: A Successful Grower's Handbook for Small-Scale Organic Farming*, the practice originated in subsistence cultures around the world—from peasant farmers' vegetable plots in the Middle Ages to nineteenth-century Māori gardeners who traded potatoes with European whalers. Fortier calls the approach "six-figure farming" for its high-profit potential—an alluring concept, especially in the United States, where more than half of all farmers currently hold down a second job. But there's a more urgent issue driving the micro-farming trend for young farmers like Carys and Benji: the growing problem of land access.

While nearly half of American farmland will be up for grabs in the coming decade, that land is increasingly unaffordable: Adjusted for inflation, the value of farm real estate (a measurement that includes farmland and the buildings on it) doubled between 2000 and 2020; the average cost of an acre is now $3,800. That may sound modest compared to the cost of, say, a house, but consider that the average American farm is 463 acres—an untenable investment for most new farmers. Then there's the risk involved with vegetable farming, since the US government, through a now $1.5 trillion, five-year leviathan of legislation known as the Farm Bill, prefers to instead subsidize junk-food commodities, such as corn and soy, and the reality that the margins can be slim. "It would be really hard to buy any amount of property just selling vegetables," Carys divulged in her homey twang. Most farmers they knew with land "either had land in the family or they're selling weed also," she chuckled.

So when Carys and Benji began having farming dreams after college, they set out to gain hard skills before pursuing their own land. They headed to California—first Baja, to work with the late Gabriel Howearth on his ten-acre permaculture food forest—then to Northern California, where Benji landed a farming internship at the eighty-acre Occidental Arts & Ecology Center. Carys apprenticed nearby with the resident carpenter at the former organization Leap Now, which had run a gap year program she undertook after high school.

The couple had first met at a coffee shop in their college town of Ashland, Oregon. Fittingly for their fun-loving farm ethos I would come to know, "I had a drawn-on mustache the first time I met him," Carys recounted as we pulled down masses of spent pea vines from the hoop house. A month after my frigid first day, it was now the second consecutive week of near-hundred-degree heat, and her green eyes flashed against her dirt-streaked face as she searched her memory. "It was early morning, I was with my [also fake-mustached] friend, we were driving to Portland, we stopped for coffee . . ." Benji told me later that Carys's playfulness caught his eye, and he asked her to a potluck that night at his house. But since she was on her way upstate, she turned him down.

Carys hadn't connected the dots, but she had heard about Benji—or rather, the legendary house he rented—since transferring to Ashland's Southern Oregon University the year before. "I was looking for communal housing, and everyone was like, *Go to Benji's house!*" Carys said. She hadn't seen it yet but had recently befriended a young woman who lived there. "I met her because she carried a ferret in her backpack," Carys laughed. "I really liked her."

This ferret-toting friend was just the sort of free spirit who fit into Benji's house scene of renegade radicals. One housemate brought home roadkill deer and brain-tanned (exactly what it sounds like) the hides to make moccasins. Another friend collected acorns around town and processed them into flour for pancakes. There was a big garden and chickens. Everybody played music. "There were only two bedrooms in this house. But there were usually five or six people living there," deadpanned Benji. He himself paid $75 a month to pitch a tarp in a corner of the backyard when he first arrived, sleeping on a bed of straw. "I was happier than I'd ever been," he said.

Benji hadn't aspired to this back-to-the-land lifestyle nor his future in farming. He had grown up in the rustic town of Sisters (population 2,781), where his parents had owned Apple Jacks, a natural foods store, before their divorce. He couldn't wait to get away from small-town life to pursue jazz guitar at Portland State University. But once there, a class with the late permaculturalist Toby Hemenway awakened Benji to the crises of our environment and food system, and he became disillusioned with the pointlessness of "locking myself in this practice room for five hours a day." He started volunteering on an urban farm and followed a girlfriend down to Ashland to pursue a degree in environmental studies at Southern Oregon. "Immediately, I found my people," he recalled. "There was this great sense of community."

Carys, for her part, was first drawn to sustainable living her junior year of high school, when she ruptured her spleen snowboarding, "had a breakdown," and reexamined her persona as a cheerleader in provincial New Hampshire. She threw herself into AP classes and set off on a gap year, which included working on a reforestation project in Tamil Nadu, India, and living in an expat organic farming community, where she subsisted on beets and observed people drinking their own pee. Urophagia aside, the experience would ultimately lead her to enroll in the same environmental studies program that Benji had at Southern Oregon. "But I still can't really eat beets," she quipped. (This aversion didn't stop Mahonia Gardens from growing the most magnificent beets I've ever seen.)

Soon after Carys and Benji's coffee shop meet-cute, Joel Salatin, the self-described "Christian libertarian environmentalist capitalist lunatic farmer" of *The Omnivore's Dilemma* renown, came to Ashland to speak. To Southern Oregon's nonconformist crowd, it may as well have been the Beatles. Amid a throng of mutual friends, Benji and Carys struck up a conversation, and she finally realized this was Benji of communal farmhouse fame. The next day, they went for a bike ride together, "and we've just been together ever since," Carys reminisced.

⁓

OUTSIDERS OFTEN IMAGINE OREGON in its Pacific Northwest glory, with verdant forests and epic rainfall. But east of the Cascade Mountains, more than a quarter of the state is known as Oregon's high desert. Technically it receives enough rainfall to classify it as steppe. The sign that welcomes you to Sisters proclaims THE WEST AT ITS BEST, and it is: The drier climate, with its 162 days of sunshine, is what makes Central Oregon a mecca for the outdoorsy—as do its epic mountains, millions of acres of forested wilderness, lakes and rivers, and endless opportunities for hiking, biking, skiing, camping, kayaking, or lazily floating down a river in summer, beer in hand. But the parched climate is also what makes the region prone to catastrophic wildfires, and that risk is accelerating as global warming drives aridification and rising temperatures in the West.

A decade ago, before Instagram and Airbnb, Central Oregon was "undiscovered" to those outside the state. But the aforementioned benefits of the region, plus spiraling housing costs in California and the exodus from big cities during the COVID-19 pandemic, have sent a swell of newcomers to Bend, Sisters, and surrounds. Unfortunately, this migration has sent housing prices skyrocketing there too.

I was among those troublesome transplants. My husband, two young daughters, and I moved to Bend in fall 2020, fleeing not only six months of pandemic lockdown in Los Angeles but more than a decade of being car-bound, computer-bound, and running on an endless treadmill of traffic and twenty-first-century futility. We told ourselves we would ride out that one year (ha!) of the pandemic, but I wondered if we might stay longer and what our life could look like if we did.

Not long after, I worked a volunteer day on a farm organized by Central Oregon Locavore. That crisp October morning, breaking a sweat clearing a field alongside the farm's intern, Katie, who waxed dreamily of all the scrumptious meals she ate from the farm's harvests, I was bitten by the farming bug. Farming, as it turned out, was the perfect career no one had ever mentioned for a foodie like me who loves the outdoors and manual labor. Katie told me she had gotten the gig through Rogue Farm Corps, an Oregon organization training the next generation of sustainable farmers and ranchers; I instantly knew I had to apply. I had long wanted the know-how to grow my own food, and had been searching for the elusive feeling of working the soil with my hands and for a life surrounded by nature—especially for our daughters. But there was something else I was longing for, though I couldn't name it yet. It wasn't until I came to Mahonia Gardens, then a host farm for Rogue Farm Corps interns, that I realized the missing component.

⁂

WHEN BENJI AND CARYS were at last ready to find land, they didn't first consider Central Oregon, where water is scarce, the growing season is short, and frost is possible any day of the year. (Many mornings washing greens in the pack shed, my arms would freeze numb from fingertip to bicep. In mid-*June*.) But then an opportunity arose in Sisters that two start-up farmers couldn't ignore: The family of Benji's high school friend Audrey Tehan offered to lease them an acre of land, at first in exchange for help around the property. Audrey would start her own educational farm, Seed to Table, on an adjacent acre that year.

So Carys and Benji mulled over the small town with the challenging climate that Benji as a teenager couldn't wait to escape— where, even when Benji was in high school, locals would muse, *Oh, I remember you running around Apple Jacks, with your curly red hair.* There had long been ranches in the region, but the market *was* undertapped for sustainably homegrown produce. Nearby Bend's population had been growing steadily since Benji's childhood, and tourists and day-trippers flocked to the western-facaded hamlet of Sisters as their gateway to the outdoors. There were coffee shops, bakeries, a drive-in with burgers and homemade ice cream, even a thriving music scene. The artsy little town had been cool, after all.

But most important, in Sisters, they would have the support of family and community. Benji's mom and brother lived in Bend, and Benji's dad, Jack Nagel, was renting a 1940s bungalow less than a mile from the Tehans' land. So Carys and Benji moved in with Jack, put together a Kickstarter, and raised $9,000—enough for a greenhouse, tools, soil amendments, and fencing. "I'm so embarrassed to watch it now," Benji admitted, then mimicked his starry-eyed West Coast drawl in the Kickstarter. *We're gonna grow seeds and herbs. It's gonna be a botanical garden and we're gonna do permaculture.* He laughed. "It just sounds so idealistic to me, but you can see exactly where we were when we started." They got jobs at a local restaurant, The Open Door, where they would work for three years while they got the farm up and running, learning about market gardening via YouTube videos. Not long after, they secured a private loan from family friends and purchased the house. And so, like the Oregon grape, the plant for which Mahonia Gardens was botanically named, they put down permanent roots.

That first year on the farm, Carys and Benji made $800. The second year, a plane carrying fire retardant accidentally dropped its load over Mahonia on the way to an area forest fire. "The entire farm was covered in red dots—the picnic tables, the hoop houses, the crops, the chard," sighed Carys. (Luckily, the FDA considers the main component, ammonium phosphate, safe to ingest in small amounts, and Mahonia's community-supported agriculture [CSA] members were understanding.) The third year, some student volunteers "weeded out" the echinacea one year shy of its *four-year* maturity, and Carys—salt-of-the-earth Carys, who greeted me every day on the farm with a hug but also sledgehammered in fence posts through her pregnancy's second trimester—made those kids dig through the compost pile to find and replant the precious perennials, then cried.

The fourth year on the farm, Benji's mom finally stopped sending him postings for more stable jobs.

When I heard these stories in the farm's ninth season, harvesting (OK, snacking on) ambrosial cherry tomatoes in a white hoop house no longer harboring red dots, it was hard to envision the tumult of those early years. Mahonia Gardens seemed to be humming toward small-farm-life utopia. With her carpentry skills, Carys had recently built "The Stand" at their house, an honor-system farm stand stocked with Mahonia veggies and staples from local producers that became a lifeline for the community during the pandemic's supply chain shortages. ("People are so honored and empowered being trusted," Carys said.) It also became a steady additional income stream for the farm: Benji was thrilled they were making as much money "as teachers." The farm now had additional gardens on the property of Carys's mom, Maribeth Quinn ("Quinn"), a nurse-midwife who moved from New Hampshire to just outside town. And most important, Benji and Carys had the support of alloparents. "Mima," as Quinn was known to the family, took turns with the other grandparents watching Junius now that he was an active three-year-old and had outgrown being toted on the farm in a baby carrier.

Carys and Benji had a successful business, they were bolstered by family, and—refreshingly—they decided to prioritize the latter. Unlike other young farming couples they knew who were forgoing parenthood to focus on their farm's revenue, they set eight-hour workdays with a half day on Fridays that left weekends free for family camping trips, Benji's local music gigs, and rest. (Aptly for country life, Benji now played dobro.) "My main goal now is to be a good dad, and in order to do that, I need to not be super stressed all the time," Benji said.

Meanwhile, when we arrived in Oregon, my family of four was living in a tract house rental supplied with Target furniture, the remnants of our LA life stacked in boxes in the garage. My kids were on Zoom school. We were relishing our nature-filled surroundings and also grateful to be alive after recovering from an early case of COVID, but this was our eighth rental since the kids had been born. We knew no one in Bend. Our family was scattered all over the country, and it would be two years before we would see them again. It wasn't until I saw Benji and Carys's homestead that I realized how sterile—how twenty-first century, how *lonely*—our life was by comparison.

Down the road from the farm, Benji and Carys had cultivated a charming compound on their own quarter acre. There was The Stand in front, their house plus a garage turned tiny home for Jack, two "shabins" where the farm interns lived from April through October, an outdoor kitchen, a greenhouse for seed propagation, a geodesic garden dome, chickens, and a sandbox and play area for little Juni— all creatively zoned, to be sure, but undeniably cheerful. It was home. Now *this is family life*, I thought.

In my quest to bring my family to live with nature and connect to our food, I had forgotten an essential part of the equation: that throughout human history, neither was possible in the absence of community—from the first hunter-gatherer bands to the later tribe, village, multigenerational extended family, or greater culture. (The outlying blip being the rugged individualism that led to America's genocidal westward expansion, the glorification of capitalism, and the preponderance of isolated nuclear families, but I digress.) A food culture, in particular, evolved from "a population's collective wisdom about the plants and animals that grow in a place, and the complex ways of rendering them tasty," Barbara Kingsolver writes in *Animal, Vegetable, Miracle*, a seminal book for Benji and Carys during their early farming journey. "It arises out of a place, a soil, a climate, a history, a temperament, a collective sense of belonging." An undeniable factor in America's deficient food culture is our impaired sense of place: Americans move more frequently than any other people in the world, over eleven moves in a lifetime. My young family and our many rentals weren't far off from this.

When I met the farm's crew, I realized I wasn't the only one adrift. Emily ("Em") Blood, twenty-nine, had carved her own path away from family in Maine and first came to Mahonia Gardens as an intern living in a refurbished school bus. She was now a beloved member of the farm, working her third season, but would leave that fall to pioneer her own farm, Sonder Farmstead, with her partner in Washington State. Cedar,[1] age thirty, one of the full-season interns, grew up on her mom's hobby farm in Oregon but was charting the wanderlust path of a Jack London in the making, including a treacherous stint as one of two women in a remote Alaska cannery. She was contemplating a farming internship in Sweden after her time at Mahonia. The other full-season intern, Elliott Blackwell, twenty-two, came into adulthood in the storied intentional community The Lorax Manner in Eugene, Oregon. He had discovered communal and agrarian life as the balm for a troubled childhood and was planning with friends to launch an ecovillage of their own.

Among this crew I had been nicknamed Farmer Jenny, and we were the extended farm family to Benji and Carys, as was their constellation of friends who would volunteer in exchange for veggies. All who worked on the farm were unflappably energetic, earnest, and happy to laugh and talk, especially about food—whether it was the carrot cake Cedar was baking, the curries Elliott was cooking, Em's and my New Englander love of sardines, or Carys's pregnancy cravings for raw rhubarb, which she chomped on like peppermint sticks when it was ready for harvest in June. Farmers, I have noticed, are often foodies, but unlike the mindless snacking of a desk job, on the farm, you are delightfully *hungry* from all that physical labor

---

1 Name has been changed to protect privacy.

when lunchtime arrives. In addition to our fondness for food, the crew also exchanged past adventures, family dramas, dreams for the future, and everything that comes up side by side while sifting through the soil for bushels of potatoes. "With farm life, you're there forty hours a week with people, so you really see their real personalities," Carys said.

Of course, other seasons at Mahonia Gardens hadn't always seen such an amiable community. You can grow a lot of food on one acre, but there's not much room to get away from someone you don't get along with. Whereas our crew would harvest in the mornings as a team, powered by Em and Elliott's silly songs, Benji's corny jokes, and communally savored coffee, the interns a prior year had stuck in their earbuds and headed out to the field in opposite directions.

And it's not just relationship issues that can dissuade wannabe farmers from life in a farming community. Carys and Benji tried to warn potential interns that working the land could be hard and monotonous. I've always relished the meditative state of working hands and body, but one young intern complained of back pain and boredom when most of the big farming lessons ended after the first two months. Midway through the season, she groaned, "Oh, we're harvesting salad *again*?" until Carys, in her warm but forthright way, reminded her: "We're going to harvest a hundred pounds of salad a day for the rest of the summer." That intern left the season early.

Communal living has its challenges too. "It's fun to be able to go into the backyard and see Elliott juggling, and just talk to him for a little bit," Benji said. Still, as Carys explained before I became their intern, "We're both a little introverted and *want* to have some alone time." The couple had come a long way since Carys's mother, before she moved to Oregon, flew cross-country to help renovate the detached garage into a tiny house for Jack. That way, her pregnant daughter and Benji could have a house of their own and a door that closes. "Sometimes you just need your mom to come out and say,

'OK, we're gonna get this done,' " Quinn told me across a carrot bed we were clearing. But Carys and Benji are always refining how to balance their needs with community life. They had set clear rules to ensure interns didn't wander into their house to put up laundry at ten o'clock at night. They also work hard to not overextend the grandparents with childcare.

I marveled at the life they had built but also wondered if I could handle such a forgoing of privacy. One day I asked Benji flat out: Did they ever have moments of *What the hell did we do?* He paused in his thoughtful way, then answered unequivocally. "I really love what we do. . . . I feel so good about how it's all worked out. Maybe I have moments of *Wow, I live back in my hometown. Am I ever gonna leave?* But that's a very rare moment. It feels good to be embedded in community."

MY LAST FRIDAY AT the farm, we wrapped harvesting at nine-thirty in the morning for a survey and first pick of the new veggies coming in. Unlike the rest of the crew, who would experience the farm's life cycle through the final layer of compost spread for November's sleeping beds, my time at Mahonia Gardens had been cut short. My husband was offered a project back in Los Angeles, and so, in mid-July 2021, at the height of summer's bounty and my daughters' frolicking time in rivers, we were heading back to the city from which we had plotted our escape a year before. I was looking forward to seeing familiar faces and places but also dreading the loss of a newfound livelihood that had felt so pure—working together for the straightforward purpose of feeding our family, the community, and the joy of creating something inherently human and beautiful.

I trailed Benji in his yellow El Sancho taco shop trucker hat and regarded the incoming harvests I would never savor: the yellow crookneck, striped Costata Romanesco, and flower-shaped pat-typan squashes glancing out from broad, spiky leaves; the fledgling jet globes of eggplant; and, in my mind, the corn and watermelon back at Mima's. We ducked into the hoop house to sample the new shishito peppers and a squatter variety called a Padrón. Later that evening, I sautéed up a precious half pint of them with olive oil and a sprinkle of kosher salt, and my ten-year-old and I ate them hot from the pan, twisting their tender bodies off the stems with our teeth and popping them whole into our mouths.

For two glorious months, we had experienced eating nearly entirely from the farm like this—from the farmer snack of sweet, just-picked whole carrots dipped in peanut butter to the bags brim-ming with veggies I took home from The Stand after my shifts at the urging of Carys: "The best part of working on a farm is getting to eat!" she exclaimed in a stylish denim jumper smoothed over her svelte pregnant belly. (Full-season interns received housing and a monthly stipend along with food; I lived off-site and was paid a sti-pend[2] and unlimited veggies.)

I had always shopped at farmers' markets in LA, but the taste of eating *this* locally was an entirely new experience. Given that land averages over $2 million *an acre* in LA, "local" farms are typically hours away. Judging by the success of The Stand, the folks in Sis-ters agreed about the incomparable flavor of hyperlocal food. Yet while Carys and Benji were proud of the sustainable food they were providing for their community and the small farm culture they were reviving, Carys wondered aloud to me one day about the lim-its of their efforts in the midst of America's industrialized eating

---

2 Carys and Benji would not take me on as an intern without compensation. "It feels good for us to be able to pay you for the work you do," they insisted. We agreed that I would donate the stipend upon publication of this book so I could avoid a conflict of interest. Journalistic ethics discourage paying or being paid by sources.

disorder. She recounted how, while working in the field, she once watched a woman pull up in an SUV to one of the new development houses flanking the edge of the farm's driveway, where the woman unpacked a trunk loaded with plastic bag after plastic bag of processed food from Costco. "Nothing in there could have even *come* from a farm," Carys moaned.

I had recently received a similarly depressing slap of reality when I checked out the newly opened 84,000-square-foot WinCo Foods chain everyone in our Bend neighborhood was talking about. I felt like I had an out-of-body experience, watching dazed customers pile car-sized shopping carts high with seemingly every processed food ever invented. Even if one shopped at WinCo to buy real food—which I will define here à la famed food journalist Michael Pollan, as food that your great-great-grandmother would recognize—how could a market garden compete with the agro-industrial subsidies that enable a $2.98 ten-pound bag of potatoes or a $15 ten-pound tube of ground beef?

After all, I knew what it meant to depend on cheap food: My brother and I were raised by a single mother, and frozen ravioli, ground chuck, and sweet potatoes meant we always had full bellies. Growing up, even Benji's family relied on Costco after his parents' natural foods store went under and they struggled with money. "My mom would bring home flats of those Costco muffins. When I was a hungry teenager, I would just devour, like, four," Benji reminisced while I knowingly laughed. We didn't have Costco near us, but I imagine Stop & Shop sourced its version from the same giant muffin factory that also fed my adolescence.

Still, Benji envisioned a day where a hundred micro-farms like Mahonia Gardens could coexist in the area, even supplying those big box stores. Indeed, this is already happening elsewhere through the growing model of food hubs, which aggregate local and regional food for larger suppliers. But with the population upswing, the demand for local food in the area was growing so fast that Mahonia Gardens and other small farms couldn't yet scale to meet it. "If we're talking about the way forward to solve a lot of our environmental issues, food justice issues, and food system issues, . . . I think small farms are a big part of the answer," he said.

Back at the farm, our pepper sampling was suddenly interrupted by the sight of Junius sprinting into the hoop house, his little blond rattail waving out of the back of his faded Teenage Mutant Ninja Turtles hat. He leaped into Benji's arms for a bear hug.

Juni reached toward the plants we had all been observing. "Ooh, basil!" he squeaked.

"Actually, it's Padrón peppers," Benji corrected in the gentle tone he used with the interns when our radish bunches didn't match or we picked the wrong-sized box for packing heads of cabbage. I glanced at the peppers' glossy, pointed leaves, nearly indistinguishable from Thai basil. The three-year-old's plant ID had been mistaken, but it was an error any basil-enthusiast adult might have made.

"Ooh! Cherry tomatoes! Do you want one?" Juni urged Elliott, then plopped a yellow jewel in the palm of his hand, followed by one in Cedar's. A few minutes later, he was crunching on a purple bell pepper like it was an apple, then ran up to me to offer a bite. "Mmm," I said.

"It's sweet!" he proclaimed, before taking off to follow a grasshopper.

That weekend, the Grandview Fire would ignite just outside Sisters. By Wednesday, my last day on the farm, the 4,500-acre fire was tearing through the forest five miles away. Mahonia Gardens, thankfully, would emerge unharmed by this disaster, but another blaze, the Bootleg Fire, raged in Southern Oregon for weeks, and it seemed the entire West was burning on my family's drive back to LA. The pandemic, of course, was everywhere. Across the world, Jakarta was sinking. While I held on to Benji's hope for more small farmers, it was hard to remain optimistic about the future in the face of all this global upheaval. Even if I decided to focus on growing food myself, where would it be safe to put down roots and call home?

Two weeks before I left, I had spent the afternoon with Carys at Mima's house, pulling the first rows of garlic bulbs from the ground before the later all-out garlic harvest. The first brush fire of the season had broken out, and air tankers were circling overhead. I asked Carys if they ever thought about moving back to the Northeast, where she was from, and where it was ostensibly safer. Her emerald eyes didn't even blink. "No."

I nodded, then offered, "Ticks."

"Ticks," she echoed. "Mosquitos. Humidity. That person who I was in high school."

We both laughed.

Months later, back in LA, shuttling our girls to and from school and writing this chapter with an endless cacophony of leaf blowers, sirens, and airplanes in the background, I thought every day about the farm. I thought about how alive I had felt, connected to the land and to community, and I knew what Carys had meant: Once you know who you are and where you belong, there's no going back. *Where* in this case might not be Oregon, at least for now, but Benji and Carys had inspired me to reclaim a new way of life for our family.

I thought, then, about Benji and Carys's new baby daughter, Lupine. She was named for an herbaceous plant in the pea family with a flowering purple spike I just the other day spotted growing wild in untended slivers of our LA neighborhood. Regeneration was possible anywhere, life was renewing after two years of the pandemic, and if I closed my eyes I could picture my journey to find such stirrings putting forth all over the country.

TWO

Wild Abundance

"Doesn't everything die at last, and too soon?
Tell me, what is it you plan to do
with your one wild and precious life?"

—Mary Oliver, *"The Summer Day"*

THE OCTOBER AFTERNOON BEFORE I FLEW TO WILD ABUNDANCE, a homesteading school tucked in the Blue Ridge Mountains outside Asheville, North Carolina, I trudged through Los Angeles traffic and took mental snapshots: a homeless encampment under a billboard touting a luxury cannabis lounge; a delivery robot tripping a teenager on the sidewalk; a man waving a giant cross in the middle of Santa Monica Boulevard. Twelve hours later, my red-eye flight was coming in for landing over Appalachia's blazing red and yellow fall.

I had come to Wild Abundance to meet Natalie Bogwalker, the school's director and founder, to learn about her classes in off-grid living—including permaculture and gardening, tiny house building, wildcrafting, and even hide tanning. But let's be honest: Given the scene I had left behind, I was driven mostly by the opportunity to temporarily escape into Natalie's homesteading fantasy and even planned a dream adventure of my own. I would take a four-day women's carpentry class and report on the role "farmsteads" could play in transforming our food system—all while camping at Wild Abundance in a hand-built pagoda in the woods.

In the small town of Weaverville, I turned off the country highway near a Baptist church and crawled down a gravel road past barns so pastoral they were cliché. I parked on the edge of a stream bank, then walked toward Natalie's cabin on a hill overlooking a paradise of green. The carpentry class had started at nine, and it was now past eleven. Natalie was nowhere to be found, and the class had already broken off into groups.

Ahead, several women sat on folding chairs under a wood pavilion, whittling points on carpenter pencils. The instructor, Ella, a young woman with a long ponytail and wrist tattoo, welcomed me into the group, whose members ranged from a twentysomething in Converse to a grandmother with a gray bob and glasses. As Ella talked about speed squares and the imperial measurement system, I glanced down at the afternoon's agenda: learning how to wield three kinds of motorized saws. *Perfect timing*, I thought, feeling the wooziness of my red-eye setting in.

Just then, a text came in from Natalie about my interview with her permaculture apprentices: *Let's all check in together at 1:07 in the garden.* 1:07? I shrugged at the probable typo, but headed there after lunch at my precise call time, grateful to miss saw class and possibly end the day with my fingers intact.

Natalie's garden was so seamless with the landscape, it was almost hard to find. Unlike the neat rows on other micro-farms, mammoth squash leaves were sprawled among grass and dandelions. A vaguely delineated bed of robust collards was strewn with straw. Bamboo trellis poles poked up everywhere, mimicking spindly trees. If I hadn't known better, I would have said it looked messy. But permaculture is a "new but old" approach to growing food that mimics patterns in nature and draws from the practice of ancestral cultures that *tended* more than *tamed* their landscapes.[1] Natalie walked into the scene wearing brown leggings, brown muck boots, and a sage green tank top. With her tangled brown curls and heart-shaped face, the effect was modern mythological wood nymph.

This was my first time meeting Natalie in person, though we had

---

[1] For this reason, permaculture (an amalgam of *permanent* and *agriculture* coined in 1978 by biologist Bill Mollison and environmental designer David Holmgren) has come under fire as of late for cultural appropriation, though much of the introspection is happening within the movement itself.

known each other since 2019, when I interviewed her for *Uncivilize*. In our phone conversations and correspondence since, she had always been warm and open. Now, as twenty-two students plus instructors milled about the open-air kitchen behind us, Natalie greeted me cordially, but I could sense a wall. It threw me.

"I love your garden," I stammered, then gestured to the giant, tooth-edged leaves around me. "Your parsnips look great."

"Those are daikon," she said.

When after a minute the permaculture apprentices didn't show for our scheduled interview, Natalie headed back toward her cabin. I decided to kill time by heading over to chop saw class at the nearby shed.

A man named Adam wearing corduroys and a pink shirt that said PAW SQUAD demonstrated how to ready our bodies before bringing down the saw. He took swigs from a giant bottle of kombucha while supervising our cuts. The women started out hesitant but soon gleefully advanced to slicing board after board, riding the adrenaline of the machine's power. As I lined up my fourth board under the blade, Natalie appeared at my shoulder. She looked annoyed.

"Make sure you're just cutting off little slivers of wood so we don't end up with lots of little boards," she told us, then walked back to the garden.

Adam nodded, then started schooling us on the rip-cut. *Oh geez*, I thought. More wasted little boards, only sheared lengthwise. Natalie hollered over to him to halt the evidently unplanned lesson.

"You can be done! You're good, bud!"

My group headed to the pavilion for the circular saw class. Natalie waved me down to the garden to finally meet her permaculture apprentices.

The three apprentices were laughing so hard that one of them snorted, clearly pals after working together at Wild Abundance one day a week for the past eight months. The permaculture apprenticeship, which focuses on growing and using perennial food crops like fruit and nut trees, was almost at its end. That bracing autumn morning, the apprentices made chestnut flour out of chestnuts they harvested, shelled, and dried in the weeks before. Genevieve Barber, a fit fortysomething blonde, stepped aside from the group to chat with me while Natalie conversed with the others before heading back toward the shed.

"Permaculture is the way we're gonna survive," Genevieve said. She owned a virtual assistant company for real estate agents, which had enabled her and her wife (who also works remotely) to recently move from Atlanta to be in the mountains. They bought a house in West Asheville with only a one-fifth-acre lot, but planned to use intensive techniques, such as the stacking functions and vertical growing she had learned in the apprenticeship.

"We're going to try to grow a ton of perennials," she said.

In the distance, I heard a howl like a gibbon in the zoo.

Genevieve and I stared at each other. The sound moved closer.

Out of nowhere, Natalie sprinted by, yelping "Ow!"

At least three yellow jackets were crawling on her body. She tried to brush them off.

Another apprentice, Margaret Clawson, walked over. She was a mom with three teenage boys and a master's in public health, and she acted like this was an everyday occurrence.

"You OK?" she asked Natalie.

"Maybe I should go the other way," Natalie wavered. She darted around Margaret, then jogged toward the other end of the garden, attempting to lose the insects. "Uhh!"

"There's one on your back shoulder!" Genevieve called out. "Your left shoulder! Your left shoulder!" She looked at me. "Shit."

Ten feet away, Natalie flung off her last wasp attacker.

"Are you OK?" I winced.

"Yeah. I just need a minute," she sighed, then walked over to a patch of plantain to make a spit poultice.

At five o'clock, the campus emptied out and the light grew dim. I was tired and my hands were numb with cold, so I decided to locate the pagoda Natalie had offered me through the weekend. I walked up into the woods, past the outhouse, and spied it. A dirty towel was slung out front. I climbed the hut's ladder to an unmade futon inside.

As if she had ESP, Natalie texted me. "I asked the cleaner to clean it, but if it seems like it does not have fresh sheets let me know and I will bring some."

I eyed the structure's woven half-walls. Natalie had gently advised on the phone that the hut was open air, but back in Los Angeles, I was envisioning stargazing. This would be magnificent in the summer, but temperatures that night were supposed to hit freezing.

I'm so sorry, I wrote back, and added that my gear (OK, fine, *I*) would be better set up for writing at a hotel in town.

So much for being an intrepid reporter, I thought. So much for my dreams of nature and self-sufficiency. I had overextended. But from the looks of the day, so had Natalie.

Was this what Natalie, survivalist extraordinaire, had imagined when she built Wild Abundance—scurrying around for sheets for a bothersome journalist, insect attackers, and her homestead paradise inundated with students? My carpentry class was only one of thirty-plus classes Natalie had hosted on her campus *this year.*

Yet Natalie was no pretentious radical, I would learn. If we're serious about food security—and if we want to chart a path to a meaningful existence instead of the automaton one civilization is marching toward—then it's time we loosen our grip on self-sufficiency fantasyland. Natalie had long since left utopia for a more uplifting vision.

WHEN NATALIE WAS TWENTY-THREE years old and newly transplanted to the Southeast, she emailed all her friends, asking for help finding an unusual living situation: She wanted to live in an intentional community, and she wanted to find one that was into eating wild foods. *Did anyone know of one in her area?*

The question wasn't crazy for anyone who knew Natalie. She had loved plants since she was tiny, growing up on the rural Washington State outskirts of suburban Portland, where her mom taught her botany and basket weaving, and took her walking in the woods. "I got into wild foods when I was like, seven," Natalie pronounced. "I remember [my mom] feeding me daylily flowers." [2] As for communal living, Natalie first discovered it in a vegetarian student co-op in college. Later, she lived at the Can Masdeu squat, an anarchist utopian community with communal vegetable gardens, founded on the site of an abandoned leper hospital in Barcelona.[3]

The spring before she moved to the Southeast, Natalie had been working as a hotel gardener in Guatemala when her father's "colon exploded" on a fishing trip and he nearly died. She uprooted to Atlanta to help her mother care and cook for her father, and by the time he recovered, Natalie—who in between world travels studied ecological agriculture at Washington's The Evergreen State

---

2 Daylilies grow wild throughout the United States, and every part of the plant is edible. A staple in Asian cuisine, they're even grown on farms across Taiwan and China. But consult a guide before foraging your own.

3 Can Masdeu became famous in 2002 during a three-day showdown between its activists and the Spanish police, who tried to evict the squatters. The squatters resisted by chaining themselves to the building, balancing on planks threaded through windows, and employing other life-threatening techniques before a judge ordered the police to withdraw. Natalie was there the year before.

College—fell in love with her new bioregion and decided to stay. "The biodiversity of the South just really did it for me," she said.

Locating a wild foods commune wouldn't be all that easy today; but this was in 2002, before those pursuits on their own had become trendy, before Michelin-starred restaurants employed foragers and one could find back-to-the-landers on Instagram, or even on a website.[4] From Natalie's mass email came a single lead: A new community called Wild Roots had just been started by a couple named James and Didi on thirty acres of inherited wilderness in the Appalachians of western North Carolina.

Wild Roots would be all about human rewilding, then a nascent movement. (Think the History Channel's *Alone*, except with people.)

So Natalie went with James and Didi to her first primitive skills gathering and then on a wild foods hike that Wild Roots was hosting with ethnobotanist Frank Cook, who the year prior had walked across North Carolina while subsisting on foraged plants.[5] "I was like, *OK, I'm moving here*," Natalie chuckled. "And so I did."

"People think North Carolina is the South and it's warm, but it gets down to like, negative five degrees here," Natalie said. That first subzero winter, she lived in a tent. Through the next, she lived in a canvas lean-to, then took a few natural building classes and built herself a bark lodge and, ultimately, an eight-by-twelve cabin made of clay and straw. She stayed warm the ancient way—with fire, started by friction—and clothed herself in deer hides she learned to brain-

---

4 I should say trendy in the developed world. For nearly a billion people worldwide, timeworn practices such as hunting, fishing, and consuming wild plant species (along with communal housing, land, and other resources) not only never went out of style, they're critical to survival—especially in the face of habitat destruction and species extinction wrought by industrial agriculture.

5 Frank would become Natalie's teacher. He died tragically at the age of forty-seven of neurocysticercosis, which may have been caused by contracting the larval tapeworm parasite on one of his overseas botanical expeditions.

tan and stitch into buckskin clothing. "Buckskin is the softest material *ever*," Natalie enthused. "It makes silk seem shabby."

Natalie drank from a stream and—as she had hoped for—had her fill of wild food. "Every morning, I would go on a two-hour walk in the woods and forage. I foraged a *lot* of food and so did the other folks who lived there . . . nettles, spiderwort, acorns," she rattled off just a few of the dozens of plant species they used. Sometimes the group would forgo the Paleolithic to drive to the best chestnut and autumn olive (not actually an olive but the juicy red fruit of a shrub) gathering spots.

They grew food in gardens and also ate, "um, a lot of roadkill," said Natalie, who before then had been a vegetarian since the age of five. She learned to track and trap, but their regular roadkill haul consisted of raccoons, opossums, groundhogs, and deer. Sometimes, meat was bequeathed by sport hunters. "You don't really want to mess with the bear hunters," she intoned. "We would let them pass through, and they would often give us a whole bear." But Natalie wasn't draconian. She had friends in Asheville, a fifty-minute drive away, and she'd often spend a day and a night there each week.

Still, she lived at Wild Roots—in a hand-built dwelling, cooking roadkill over a stick-sparked fire, cloaked in buckskin—for *five years*.

Natalie, now forty-three, was telling me this story the evening of my second day at Wild Abundance. She had invited me to dinner, and we were inside the woodstove-toasty kitchen of the two-story log cabin she built by hand. She seemed relaxed in her homey surrounds and was whipping up a meal surpassing any farm-to-table restaurant: locally raised chicken roasted in cast iron with red onions and za'atar; broccoli we harvested from her garden, sautéed with garlic; an assortment of peppers from her neighbor's garden, charred on the Weber on her deck; and chestnuts from her trees, roasted until they blistered and slathered in butter. "They're little carb packets," she laughed as I *Mmm*'d at an improper volume upon my first sweet bite. "They're a really great thing for survival. For a lot of cultures all over the world, it's been their staple."

Earlier that morning, in a Q&A with the carpentry students—who had traveled from as far as Minnesota and Ontario to take the class—I had watched Natalie entertain her audience like a seasoned pro, deftly weaving in plugs for other Wild Abundance classes in between anecdotes. That afternoon, the YouTube show *Tiny House Expedition* had been there filming, the latest in a string of media types (ahem) seeking Natalie's land-based living expertise.

And still, I couldn't reconcile that the sociable storyteller—not to mention savvy businesswoman—in front of me could have lived like such a hermit. Or been so hardcore. "What do you remember most about living there?" I asked. "How did it feel?"

"It was a lot of things . . ." She searched for the right words, then whispered with reverence. "It was *so* beautiful. Just living in wild land, in western North Carolina . . . it's like the most biodiverse place in the temperate world." She crushed a clove of garlic with a mason jar.

"And I also remember being lonely because it was basically me and a couple for a lot of the time that I was there."

I raised one eyebrow and she laughed it off, conceding, "I definitely had some love affairs" with Wild Roots' ever-changing cast of visitors. But by and by, the experience proved far more *intentional* than *community*.

"And that sucks," she laughed, her forehead crinkling. "Like, who wants to live just you and a couple?"

By and by, Natalie realized possessing the skills for an apocalypse (or future pandemic) in isolation didn't feel fulfilling. "I had been . . . in total blissed-out rewilding adventure land, and I wanted to share that with people," she told me on her deck under the glow of the Milky Way. "I'm a good builder, I'm a decent gardener, and I'm good at lots of things. But what I'm really good at is inspiring people and organizing people. . . . That's really my gift."

I can concur. To some, Natalie's backstory may sound like she's out there or enjoys foraging for psychedelic mushrooms as much as she does wild greens. She doesn't. "I have an interest in all these entheogenic plants," she said after I asked about a plant in her garden with orange flowers that turned out to be dagga, acknowledging some had been helpful for expanding her mindset when she was younger.[6] "But really, I'm pretty boring. My life is so full that if I'm not *on it*, bad shit happens. And I get migraines."

In person, what is most striking about Natalie is how she delivers a message of deep nature connection from a persona that reads—sorry, I don't know how else to say it—*normal* and *cool*. I've logged time on the ancestral skills circuit, both as a reporter and an outdoors enthusiast, and I can attest there is no shortage of "characters" in that world. (Once in a backcountry shelter class, a classmate stared into my eyes and proclaimed I needed a checkup for my lymphatic system.) Natalie is not one of them.

Sometimes it seems as though even Natalie has a hard time believing she's lived the life she has. When I told her I grew up with a similar affinity to nature and asked what she thought it was about her psyche that drove her to live in a hut in the woods for five years, she exclaimed, "I would not do that now!"

After our laughter subsided, she added: "I think that I have a lot of care for the more-than-human world. And I deeply enjoy ancestral skills. It's really fun for me."

Near the end of her isolation at Wild Roots, Natalie and Didi began traveling to college campuses to give slideshow talks about primitive living. Natalie funneled her speaking fees into launching

---

6 Dagga is a shrub in the mint family that purportedly imparts a mild high from smoking the flowers, though Natalie hadn't tried it. The plant is originally from South Africa, where it is used in traditional medicine, which is confusing because the slang word *dagga* in South Africa also refers to cannabis. This species, *Leonotis leonurus*, is legal and uncontrolled in the United States (except Louisiana).

a green anarchist event called Feral Visions. By the time she thought up and organized Firefly Gathering, a multiday event turned non-profit that has become one of the largest earth skills events in the country, she realized it was time to get a computer. That year, 2007, she started sharing her skills, teaching her first hide-tanning class.

Then, in 2010, seventeen acres came up for sale behind the home of two acquaintances. Natalie pounced on the opportunity to purchase the land with a small group, paying around $50,000 for her own seven-acre parcel—the one with more woods. With the cost of land skyrocketing, Natalie encourages co-buying as a doable strategy to secure land for homesteading; commercial farmers are doing this, too, as I'll explore later. "This land was not as wild as I might have liked, but it had community going for it," Natalie said. "And it's not subdivision land, it's country"—just twenty minutes from downtown Asheville.

The open section of the land where her garden and campus are now situated was "not super usable." It was sloped, and flooded, and only about half an acre. But "you can do a lot on a half acre," she encouraged our class in the Q&A. "I went for it. And I didn't have much money." With the help of friends, Natalie would build the homestead, plant the garden, start the school, and teach the classes. As it turned out, it was Natalie's love for the wild—paired with her inborn business talents—that would lead to the unique mark she is making on the world.

※

NATALIE TOOK THE LAST name Bogwalker in her early twenties, inspired one day while gathering blueberries. But there is an alternate portal through which Natalie, whose original surname is Nicklett, might have had a conventional life. Her father is a civil engineer who managed a diaper factory, of all un-eco things, when Natalie was growing up. Her sister is a professor of public health who did her postdoctoral fellowship at Johns Hopkins. Natalie herself set out to study genetic engineering (she wanted to find the cure for AIDS; Natalie does nothing halfway) until one day, at the age of 19, she was biking the Burke-Gilman Trail in Seattle. She came to a road crossing, looked both ways, and found herself smashed against a car's windshield before being flung into a telephone pole.

She describes that accident, which led to no broken bones but which took months to recover from (yes, she was wearing a helmet), as "an initiatory experience" present in traditional cultures but now absent from our own. "That was the best thing that ever happened to me because it really shook me to my core," she said, then echoed poet Mary Oliver—"Oh, wow. I could die. What am I gonna do with this one very precious life?"

So Natalie took a break from school and, aided by a little settlement money, backpacked around Western Europe for the first time. Then down the rabbit hole she went: She worked at environmental nonprofits, started an organic catering company, took "epic bicycle journeys," went back to school for ecological agriculture, got "tear-gassed and beaten up by cops" at the 1999 World Trade Organization anti-globalization protests, then found herself at the forefront of the green anarchist and rewilding movements. "I was very radicalized," she said about the prelude to her time at Wild Roots.

My own awakening to the illness of industrial civilization started a decade after Natalie's, but here's the difference: By that time, I was enmeshed in a paycheck-to-paycheck urban existence. I was married with two little children. I took survival skills classes, dabbled in gardening, wrote about nature, and started a podcast (just what everyone will be vying for when the grid goes down). But in September 2019, two weeks after my first interview with Natalie, I, too, was in a horrible car accident—hit by a drunk driver—and her words thereafter circled in my brain. *What are you gonna do with this one very precious life?*

Was this my initiatory experience? Honestly, I still don't know; living in Los Angeles is a daily experience of astonishing events. But I knew that until that point I had been only shuffling toward meaningful change.

Then came the morning of March 11, 2020, when I realized one hour before everyone in Los Angeles that we were about to be thrust into a global pandemic and ran to Trader Joe's to pad our emergency pantry. As I tossed bricks of frozen beef and bags of wild rice into my cart, an image of Natalie popped into a thought bubble above my head. I imagined her gathering early spring greens and sorting seeds to plant her annual vegetables. My eyes tracked the incoming masses, who were throwing frozen samosas and canisters of whipped cream into their baskets as levelheadedly as the passengers getting into crash position in the movie *Airplane*. I wondered how many jars of pickles and peaches Natalie had left in her pantry. Hell, she could process a deer! I feel obliged to disclaim this scene by saying I was lucky for the means to stock up on staples and the cooking know-how to make it last, but I knew my cart held maybe two months of supplies. Natalie, on the other hand, had been readying for this moment for decades.

Yet back in Weaverville, Natalie was actually scrambling, just for different reasons. Before COVID, Natalie had filmed and started promoting her first online class, a hide-tanning class. But Natalie no longer lived a purely subsistence lifestyle. Her livelihood—and the livelihood of her employees—relied on Wild Abundance's in-person classes. And her school was about to be shuttered.

"We had to do a big pivot," Natalie recalled. "COVID happened and we were like, 'Wild Abundance is gonna go under. What are we going to do?'"

So Natalie and her friend Chloe Lieberman, who homesteaded on a nearby twenty-three-acre farm and was copywriting for Wild Abundance, decided to put their energy into creating another online class—an organic gardening school.

Their timing couldn't have been more perfect. Supermarket shelves were empty. Forty percent of Americans were experiencing food insecurity, many for the first time. Home gardening supplies and seed sales were soaring. Not since the Great Depression were more Americans attempting to plant vegetables.

Natalie and Chloe launched the Online Gardening School, which consists of a library of hundreds of how-to videos encompassing their combined half century of horticultural know-how, as well as access to live Q&A calls. Unlike other online classes out there centered around ornamentals, Natalie and Chloe's focuses on growing food—including month-by-month instructions on more than twenty-five vegetables, herbs, and berries using organic techniques. You can watch a free introductory class, "Top 10 Vegetables to Plant That Will Really Feed You," before signing up for the course. "It was very popular during the pandemic," Natalie said. "Very, very popular."

Then came more luck: Natalie realized they could continue Wild Abundance's in-person classes, which were outdoors, provided everyone was masked and distanced. They only had to cancel two classes out of twenty that year. "People were *so* thankful to be able to come here and have something to do with other people. . . . It was our best year," Natalie said, echoing nearly every farmer and food activist I would interview for this book.

BY THE TIME I visited Wild Abundance in October 2022, the women's carpentry classes had more than 3,000 people on the interest list. The in-person tiny house class had more than 4,000. Yet gardening school sales had dropped off since the height of the pandemic. (Note: Wild Abundance offers a sliding scale for many of its classes, as well as scholarships for those who can't afford them.)

Chloe, now codirector of the Online Gardening School, offered one explanation: "I think the ubiquity of cheap food in the US makes self-sufficiency gardening really a different thing here than it is in other [countries] because you can just go to a big box store and buy a bunch of cheap food for way less—if you translate the money into time—than you would spend on gardening and growing that food . . . although that's changing as food gets more expensive."

Indeed, it's now a few years past the supply chain collapse of the early pandemic, but as I write this, eggs are scarce due to an avian flu outbreak. The cost of groceries is skyrocketing as a result of inflation and ongoing disruptions. This is a book about farming, yes, but the role homesteading could play in bolstering our food security should be reexamined.[7] That may sound idealistic, but already, five of every six farms in the world are under five acres, according to the Food and Agriculture Organization of the United Nations. These "smallholders" raise livestock and grow grains, tubers, pulses, vegetables, and fruit largely for their own consumption, and in total produce 35 percent of the world's food.

Granted, many of these subsistence farmers are concentrated in impoverished areas of the world, but it's evidence that home growing, collectively considered, can add up to a lot of calories.

"I think that [commercial] farming is awesome. But it's not for everyone," said Natalie, who in agricultural school thought maybe she'd be a farmer but found the lifestyle "backbreaking" when she toured farms in her program not long after her accident. She was dismayed, too, to learn how difficult it was to eke out a living in the United States as a new farmer on rented land. "In Europe, they sub-

---

7 Although perhaps it's time for a new term. Chloe pointed out that *homestead* is "a problematic word" in the United States, where it's embedded with individualism not to mention a legacy of Native land dispossession.

sidize small-scale farming. It's a hard path to make in our particular economy and culture."

Particularly, the staggering cost of health care in the United States can stall aspiring farmers before they get started. "I am really dependent on [employer-based] health insurance because of my son," said Dan Hancock, fifty-four, Natalie's third permaculture apprentice. His son was born with a heart defect and has required several surgeries. Before Dan had kids, he secured an apprenticeship at an organic farm. In the economic uncertainty after 9/11, though, he wound up taking a job with natural grocery chain Earth Fare and, later, Whole Foods, where he's been ever since. "Without [Wild Abundance], I don't think I would have gotten the jump start I needed to really dig in and feel like I can do this, I can come up with a plan." In five years, he wants to buy more land, grow and raise the majority of his family's food, live simply, and teach other people.

Now, I am not the first person to tout "farmsteading" as a means to food security or a more purposeful life—even with minimal land available like where I live. Carleen Madigan, editor of the book *The Backyard Homestead*, estimates that if you have a quarter acre—the average residential lawn size in the United States—"it's possible to produce 50 pounds of wheat, 280 pounds of pork, 120 cartons of eggs, 100 pounds of honey, 25 to 75 pounds of nuts, 600 pounds of fruit, and 2,000-plus pounds of vegetables."[8] More urgently, "gangsta gardener" Ron Finley started a movement growing fruit and vegetables in littered sidewalk parkways in LA's "food prisons" (his apt term for where systemic racism has left neighbor-

---

8 Unbelievably though, in many places it's illegal to grow food in your yard. "Right to garden" laws have recently been passed in two states, and activists are working to enact more.

hoods devoid of a grocery store), and there will be stories of activists like him in this book.

Yet what should be added to the greater conversation is the need for more schools like Natalie's. Growing food was a skill handed down from generation to generation for at least 12,000 years of human history, practiced in connection with a specific climate and landscape, individually learned through thousands of hours of observation and dirt time—yet we seem to think this lost knowledge can be reabsorbed by reading books and watching a few YouTube videos. We try our hand at gardening, fail, and throw out the oft-repeated refrain, *Oh, I don't have a green thumb*, forgetting this is an intelligence our modern culture no longer teaches. "One of my old land mates was a brilliant, brilliant man . . . an engineer," Natalie said. "Engineers, you know, know how things work. And he was like, 'Oh, I just figured it out. Fruit comes from flowers.' . . . He was forty-two years old at the time."

What we have right now is a precious spark of enthusiasm for reclaiming gardening wisdom. The National Gardening Association reported that more than 18.3 million Americans started gardening in 2021 alone. "There's a huge resurgence of people who are really excited to provide for their needs and to feel empowered," Natalie said. "People are *starving* for it, and for good reason. I think COVID really allowed for that to blossom." In other words, the pandemic was a collective initiatory experience—one that prompted so many of us to consider alternative paths for our very precious lives.

But can we restore food autonomy, specifically, in a lasting way? We don't yet have good statistics to know what percentage of growers will continue, but history may lend some foreshadowing: The Victory Garden movement that emerged during World War I (and the 1918 influenza pandemic) saw Americans grow 40 percent of the nation's fresh vegetables when it resurged in World War II. But what's seldom noted is what happened to these inexperienced gardeners after the wars were over: They threw in their trowels and went back to the grocery store.[9] "No amount of warning will make people plant their Victory Gardens again this year unless they are convinced that they are really needed," the *New York Times* reported in 1944.

But if anyone knows how to stoke a spark into a fire, it's Natalie. So toward the end of our dinner in her cabin, after I had polished off more roasted chestnuts with butter than was probably polite, I asked her: How does someone like me actually forge a radically different kind of life? And while we're at it, how do we scale an entire regenerative movement?

"I mean, there's lots of different ways that people can connect with their food system," Natalie said in her upbeat, accessible way, running through actions for conscientious consumers, including shopping at the farmers' market and switching to buying organic, sustainable, and locally grown food.

"But I think that getting your hands in the earth and growing some stuff, to me, is a very radical practice. It connects us with our ancestral lineage. It connects us with the earth. It connects us with the joy and excitement of seeing a seed you planted germinate."

Natalie was sitting across from me at her inlaid wood kitchen table with branch legs, but I could see the wave of wonder welling up inside her. "It's like, oh my gosh! I've been gardening for thirty years, which is bizarre to say," she laughed. "And when I see seeds germi-

---

9 What's also seldom noted is what really led to the rationing and food shortages that prompted the US government to propagandize World War II's Victory Gardens: the incarceration of 120,000 Japanese Americans, who until that moment constituted two-thirds of the West Coast's farmers (growing 40 percent of California's vegetables alone). This chapter in our nation's history was shameful even before the evidence emerged that internment was orchestrated by the US government as an agricultural land grab for corporate agribusinesses. Victory Gardens were the decoy.

nate, when my row of carrots is finally picking up, I'm still like, *That is so cool. This is happening! This is amazing!* And it's magic."

NATALIE IS NO LONGER the purist of her youth. As I stood in her mudroom and put on my boots to bid her good night, I eyed the woodshed just outside her door, stacked with logs toward eternity for the coming winter. "Did you chop all that wood?" I asked with astonishment. "No, I bought wood this year," she smiled. "I used to, though!" She also shared she recently had the natural wool insulation in her cabin replaced with the conventional kind. The wool was supposed to have been antimicrobial and repellant to insects, but over the years got filled with bugs.

But a bigger concession had come to a head in the years since the pandemic, when Natalie started experiencing social anxiety for the first time in her life. Instead of using medication or even natural modalities to release it, she *listened* to what the anxiety was telling her: Running the school, teaching the classes, managing the homestead, and raising her young daughter was "too much for one person to hold." So Natalie transitioned into the role of Wild Abundance's director, finding other instructors to take over her classes.

"I had a lot of my identity tied up with teaching, and it was a big deal to let that go," she said. "But it's been such a blessing." Clearly her new role fit. When I first showed up for dinner, I asked if she was tired. "It's fun!" she said, insisting that hosting a full-tilt carpentry class, YouTube crew, and a trailing journalist wasn't an everyday occurrence. "I think those stings gave me a boost."

What may have accounted for the disarray of my first day, though, was another transition underway. Natalie had recently purchased nine acres of creekside land, five miles away from her homestead, that would soon host the majority of the school's classes. This would return some privacy to her personal life as well as restore the goodwill of her land mates and homesteading neighbors who "were up in arms about the amount of traffic," she explained. Natalie was even contemplating traveling to Spain with her daughter in the spring—an extended trip away from Wild Abundance that hadn't been possible for years.

Most of all, Natalie hoped these adjustments would help spread her work to the community beyond so others could find the sense of purpose and belonging that she was once searching for. "When I was a teenager, I didn't know anyone who felt the same way that I did, and I had such strong beliefs and such a desire to connect with the more-than-human and human world." She paused, intentionality incarnate, and I could see her wheels turning to craft her message before I made my way out the door.

"I try to have a really big reach because I want people who are locked into the conventional way of living—forty hours a week of work, drive here, drive there, the way you have fun is you drink and pay a bunch of money for entertainment rather than coming up with creative things on your own, and you watch TV or play video games or get on social media—to have the experience of knowing that there are other options and that there are ways of relating with the earth, with our livelihoods, with each other, that are beyond the status quo. So I think my work for the last twenty years has been about exposing people to that and then giving them the skills to be able to create a different world for themselves."

I didn't end up going to my carpentry class the next day at Wild Abundance. I decided to fly home early to get back to my deadlines and my upcoming interview trips, but most of all my daughters. With more teachers like Natalie, they could find their way to a connected life earlier than I had, I hoped, as I headed down the stairs of Natalie's cabin. At first, I reached for my iPhone to light my way to my car, but then I realized Natalie had shown me the stars.

THREE

Ambler Farm

"It is our joy to live with less so that
others may have enough."

—Doris Janzen Longacre

I WAS INTRIGUED BY DREW DUCKWORTH'S STORY BECAUSE, BASED on his bio, he fit the stereotype of many men I had grown up around in the old-moneyed towns near New York City. Drew grew up in Rye, New York, the son of an interior decorator mother and a trust and estate lawyer father. He attended elite New England boarding schools; first, The Fessenden School, whose alumni included Howard Hughes and two Kennedys; later, The Hotchkiss School, founded in 1891 to ready young men for Yale. Six-foot-four Drew played lacrosse at Pennsylvania's Gettysburg College, studied management with an economics minor...Drew filled in the pigeonholed narrative he himself envisioned upon graduating when we first spoke: "I was like, well everybody I grew up around hops on the train in the morning, commutes to the city, and then goes to work in an office, and then comes back home and has dinner and does it again. And then they play golf on the weekends....So [I] never really questioned it and just was headed on the train that way."

Yet seemingly, Drew had transformed himself into someone unlike anyone I encountered in the *Stepford Wives* suburbia of my childhood (a culture of two-acre zoning and restricted country clubs where I grew up in an apartment). Drew, thirty-six, became a farmer. What's more, he was now the farm manager of Millstone Farm in Wilton, Connecticut, the Fairfield County town where I went to high school—a town that had no flourishing farms twenty years ago. I heard about Millstone from family who recently moved back. They also said the whole region of New York City commuter towns was teeming with young farmers and homesteaders reestablishing the area's pre-Revolutionary farming heritage. For a while, my husband and I had even toyed with the possibility of relocating there. I hoped Drew's story, and my hometown's story, might deliver a microcosmic glimpse of the regenerative farming movement in America.

And it did, although not as idyllically as first envisioned. I found preservation and perseverance but opened a Pandora's box on the accelerating inequity between wealthy landowners and cash-strapped new farmers—abetted by US economic policies that support land as an investment for the rich rather than a means to grow local food.

But before I launch into Drew's parable, I have to come clean about something. At this point in writing the book, three months after signing my contract, I thought I might have to walk away because of how hard it was proving to get in touch with farmers. Out of fifty-plus interviewees I initially reached out to, a mere three had actually responded. Later, farmers would ghost me for phone interviews. Others wrote they would only grant an interview if compensated, the first such response in my career. I knew starting a book about farming during harvest season wasn't ideal, but this was unanticipated radio silence. In my preliminary year of research at the height of the pandemic, farmers were busier than ever yet filled with purpose and enthusiastic to speak with me. Clearly, something was going on.

So when I called Drew in August 2022 for our scheduled interview, I was surprised when he picked up the phone, and I asked if I had caught him on his lunch break. "There is no lunch break," he laughed as he walked from the field to his office. Drew was similarly forthcoming when I asked about his conventional upbringing and how he came to this radical shift in his life path.

"My mom had a small garden. But I never did anything in it. Never had an animal outside of a goldfish. Disconnected from the beginning," he recalled. Drew had the resonant inflection of a finance bro if a finance bro peppered his storytelling with signaling about a spiritual awakening. "Then I went to college and chose that [school] for athletics. . . . I never really thought about *Who am I? What do I want to do? Or how do I want to be in this world?*"

A few months before Drew graduated, at the tail end of the Great Recession in 2009, the stock market crashed. "Thankfully, there were no jobs," he said with the knowing of one who ostensibly escaped his fate. "So I took that opportunity to travel." He looped across the country and into Canada, camped at national parks, did some backpacking. "And I saw that there's a different way to live your life," he said. "Basically, I got out of the bubble of Westchester and Fairfield [counties] and saw that, yeah, there's beauty and nature in the world."

Now, New York's Westchester County and Connecticut's Fairfield County are so picturesque that people flock there to visit. Drew was talking to me amid undulating countryside and seventeenth-century houses, down the road from Weir Farm National Historical Park. But in his twenties, it took Drew time to take off the blinders. Once back, "I didn't have any role models to really follow," he said. "So I started doing real estate in Manhattan."

Drew reconstructed the scene. "I was in a suit every day, walking around the city just miserable. And I found myself in Central Park or at the [American] Museum of Natural History at the end of every workday, not caring that I didn't make any sales. I was never driven by money. It was just soul-sucking. I was like, *OK, I need to do something that I want to do.*"

Traveling had rekindled Drew's memories of sleepaway camp in Maine as a kid. "I knew I loved the outdoors," he said. "So how do I get outdoors to work?" He signed up for a course with a wilderness school, National Outdoor Leadership School, followed by Outward Bound. Not long after, a family friend, an educational consultant, asked Drew if he had thought about a career in wilderness therapy.

"I researched [it] and I was like, *Wow, this seems like exactly what I want*," Drew said. "Using nature as a tool and a connection point for people to get outside of their comfort zones, go through a little darkness, and come out on the other side to some light of who they are, how they want to be in this world, and gain some skills to follow that."

I wondered if there wasn't some personal darkness he might be alluding to but decided to save that question for when I would travel to Millstone.

Drew continued the story to Colorado, where he worked for three years as a field guide for adolescent youth and outside of that work discovered another passion: wilderness rites of passage. He undertook several trips plus a nine-month experience called The Immersion at Washington's Wilderness Awareness School. "The wilderness is a great place for a mirror to be held up, and you can't point the finger at anyone else," he said. "That's just who you are."

I shelved that follow-up question too.

But as Drew segued from his passion for primitive skills into his path to farming, I detected the first blip from self-assuredness into uncertainty. "I went back to wilderness therapy for a little bit," he said, "and then I was like, *All right, I need something more stable*." He wanted to continue working outdoors, but the week-on, week-off schedule was hard for him. "Then I was like, all right, *farming*. So I just kind of fell into it."

This didn't sound like the epiphanic moment I had heard about from so many other new farmers, so I asked if there was anyone in particular who turned him on to the idea. He paused, and I heard him search his brain. "Yeah, I mean uh, Michael Pollan," he said, pronouncing the author's last name *Pole-lan*. "I remember being out in the backcountry, reading *The Omnivore's Dilemma*, and every line being like, *Yes! Yeah!*" From there, he said he "became pretty voracious," reading and watching videos about farming online.

"I also just committed to *I want to do this*. So let me go live in a barn loft, and I cut my teeth and just learned from people who are doing it."

That turned out to be the Maine Organic Farmers and Gardeners Association, through which he found an apprenticeship, then later Wild Folk Farm in Benton, Maine, where he helped pioneer a rice paddy in the state using *aigamo*, a Japanese method of rice farming that integrates ducks for pest and weed control. (Wild Folk also farms hemp for CBD; small farmers in the United States cannot earn a living on heirloom rice cultivation alone.)

Drew described those years as "simple living" in "a screen house in Maine." Juxtaposed with his current position at Millstone Farm, which he came to in 2019, he seemed a long way from that time. Millstone's seventy-one acres were restored by philanthropists Betsy and Jesse Fink with the help of "farmer to the stars" Annie Farrell. Drew's bosses were now Eliane Cordia–van Reesema, a dressage champion, and her husband, shipping magnate Volckert van Reesema. They purchased Millstone in 2016 for $5.9 million and added an equestrian facility to the agritourism site, where Drew oversaw a head gardener, head of livestock, a facilities caretaker, and a groundskeeper. "I'm more in a management position," he said, "which is a lot of office work."

Now, Drew said he was trying "to find the balance of passion and pay. . . . There are definitely some sacrifices I've made that were consciously made." He was also getting married at Millstone in early October in "a lavish, country-style ceremony," as it would later be described on the website Eurodressage. His fiancée wasn't a farmer but an equestrian and dressage champion, a Canadian Olympian. (Millstone co-owner Cordia–van Reesema introduced the couple.)

"For me, it was: How do I walk in both worlds so I can take my partner out to dinner and not feel like I just threw the seed money for the farm away?" Drew said. "While I'd love to be doing chores and seeing the animals every morning, that wasn't gonna provide me with the finances to live the life that I want."

But could he live that desired life in that exclusive area of the country on a farm manager's salary? I asked. Drew paused. "Yes," he said, but "I wouldn't say that it's super common for many farm managers." He explained that in addition to his salary, the Van Reesemas provided him with housing. "They've provided me with little gifts too." Here he chose his words carefully, but added that Millstone's owners offered him the extra position of barn manager for the equestrian operation so he could travel down to Florida with his new wife and Millstone's horses for the winter show season.

"You're not owning seventy acres in Wilton if there's not financial means, right? And they're super passionate about farming," Drew explained. "We're not making a killing here by any stretch of the imagination." Millstone's owners had the resources; he had the skills. They said, "We want farming to happen on this landscape," Drew expounded. "And we're trusting you to lead that vision."

I looked forward to reporting at Millstone that fall on the breadth of sustainable farming with full financial backing. Drew later confirmed our in-person interview for the Friday before Halloween, so I took another red-eye flight back to the East Coast.

The morning I readied for Millstone was the stuff of autumnal New England postcards. As I zipped up my down jacket over my overalls to head out the door, my back pocket buzzed with a text. It was Drew.

How long are you in CT for? Can we try to reschedule?

It was 7:44 a.m. Our interview was scheduled for 8. I called him.

What's going on?

I'm wiped, he said.

Are you sick?

Yeah, I think I'm sick.

Do you think it's COVID?

Uh, maybe.

I reiterated I was in town for one day and had flown across the country to meet him.

Could I tour the farm with someone else on your staff, perhaps? I asked.

That's not possible, he said.

I said I'd check in at midday in case he was up to briefly meeting at a distance or someone on the property could let me snap a few photos.

At noon, Drew called. *Yeah, sorry. My body is telling me I can't get out of bed.*

I wondered how Drew had handled his body's protestations on those rites of passage trips or when rising for farm chores back when he "cut his teeth" in a barn loft.

I was miffed, to say the least, but being stood up also fueled my suspicion there was a troubling reason Drew had originally departed from his privileged background to reconnect to the land, first as a wilderness guide and then as a farmer. Was it a love of nature and helping others, or simply to heal an earlier alluded to "darkness" inside of him? The latter was a common thread among new farmers, I would later notice, and I theorized it was why so many had been unwilling to speak with me: The glow of back-to-the-land living had faded; the financial realities of farming had settled in. With no higher purpose than themselves, they were stymied.

Drew didn't follow up on my request for a later phone interview (the Van Reesemas didn't respond to my request for an interview either), but here's the thing I know about affluence, having always lived on its outskirts: It enables you to prioritize *want*—a word Drew used seventeen times while recounting his story in our initial interview—rather than what those around you *need*.

The Stones put it best: You can't always get the former. But if you try sometimes, you might find you get the latter. I didn't get the interview I had hoped for at Millstone, but later that day I was the privileged one—gifted with a vision of a true, inspiring way forward.

⁂

FOUR HOURS LATER, I pulled up to a shingled gingerbread carriage house with a sunflower-boutonniered scarecrow out front. An expanse of leaf-strewn grass rolled down to a white antique farmhouse. The farmhouse was the Raymond-Ambler House, built in 1799, and I had arrived at Ambler Farm, another historic farm in Wilton revitalized since I moved away. But unlike Millstone—a private farm procured by inordinate wealth—Ambler is a community farm designed for public use and owned by the town itself.

This is a pivotal distinction. In the United States, 95 percent of farms are privately owned by individuals, family partnerships, or family corporations—together classified by the USDA as "family farms." This warm and fuzzy term is slapped on labels to connote sustainable practices but is actually a load of greenwashing, since even commodity farmers growing genetically modified soy meet the "family farmer" definition. What's more, private acquisition of land—especially farmland, for reasons we'll explore—is only increasing in avariciousness as the rich grow richer. A recent analysis by Bloomberg and the *Land Report*, a magazine for investors in "America's most valuable natural resource," revealed that the country's 100 biggest landowners own 40 million acres—equal in size to Florida. Bill Gates is now the largest owner of private farmland in America, with 242,000 acres across eighteen states. In this context, farmland owners like the Van Reesemas are small potatoes, and Ambler Farm, a community resource, is an astonishing outlier.

At the core of Ambler's community calling are its educational programs, including a summer farm camp, farming apprenticeships for school-age children, as well as workshops on garden planning and even maple sugaring. There are also myriad public events, such as its summer concert series. I arrived the afternoon of Fright Night, whereby a hundred third to fifth graders would descend on the farm for a moonlight hayride and spooky stories around the fire.

I had inadvertently crashed the pre-party: A few hours after the Millstone misfortune, Ambler Executive Director Ashley Kineon called, apologizing for not getting back to me amid the hubbub of Ambler Farm Day, their earlier fall festival. *Could I make it over in the next twenty minutes?* she asked.

So now, Ashley and I were striding uphill toward a red barn to meet Ambler's farmers in the organic production garden as approaching dusk scattered filtered light across the landscape. "You just missed Betty Ambler. She was living in this house when you left," she said cheerfully, referring to when I was in high school.

We stopped near a wall of the iconic Connecticut variety, hand-cobbled from the fieldstone agrarians unearthed from this once glaciated countryside for centuries, and Ashley sketched out the farm's history.

Elizabeth (Betty) Ambler, born in 1919, had been the last of six generations—spanning nearly 200 years—of the Raymond-Amblers to live on the farm. The family had grown subsistence crops like oats, potatoes, and vegetables throughout the nineteenth century, when farming and local industries such as shoemaking and distilling were the town's mainstays. But the rise of industrialism and urbanization saw the town depopulated by the early twentieth century, and many of the area farms deserted. Later, an influx of summer home–seekers and subdivisions for New York City commuters left the town's agricultural heritage all but erased.[1]

By the time Betty Ambler died in 1998, "It was a working farm in a sense that she produced hay," said Matt Oricchio, Ambler's property manager and assistant director of programming, who had walked up to join us and looked to be in his early thirties. "She had a couple of animals that were essentially pets."

"Everything was in terrible disrepair," Ashley added.

Wilton's citizens were worried that Ambler's land, which had dwindled from the farm's original 300 acres to thirty-one, would be lost to developers. So in May 1999, Wilton voters approved the purchase of twenty-two acres of Ambler's property as part of its open space initiative, "with the purpose of it remaining a community asset for people to use," Ashley said. In 2005, the Friends of Ambler Farm nonprofit formed to restore the salvageable buildings and develop the site's educational and agricultural vision. "So at that point . . ."

The story was interrupted by the buoyant approach of a man in mud-stained cargo pants and an ear-to-ear grin. I didn't hear what he called out because Ashley and Matt immediately burst into laughter.

"This is the farmer," Matt said.

"He's a character," smiled Ashley.

"He's wonderful," Matt added.

"Hey!" Jonathan Kirschner, forty-four, Ambler's director of agriculture, said as he strode up to greet me. I had heard from my family in Wilton how beloved "Farmer Jonathan" was in town and relayed

---

1 Of course, the area's agricultural legacy long predated the colonial era and twentieth century. The Siwanoy, Paugussett, and other Native peoples of the region selectively harvested the landscape and planted seasonal farmsteads of maize, beans, and squash for millennia.

their admiration. "It's all a lie," he said, sounding and looking like a younger Albert Brooks who had instead ventured into farming.

In keeping with that description, the main currency on the farm was undoubtedly humor. When I asked Jonathan why he started farming, he replied, "I mean, you have to want to earn no money. Live really simply. And work really, really hard. So for me, it worked out great." (Cue the rimshot.)

As a journalist, you learn to keep an open mind but, ahem, trust those first impressions. My other immediate takeaway was that this was a group who liked to have fun—no surprise for a farm that hosts 1,200 local kids throughout seven weeks of camp every summer. "When the summer kids come, they just run over everything no matter how much you tell 'em," Jonathan chuckled. "But it doesn't matter because, as a farmer, one of the things you learn is, eh—it's gonna be fine."

The fields had seen their first light frost, but Jonathan jaunted from bed to bed, displaying the remaining bounty of some forty-odd fruits and vegetables they grew for Ambler's farm stand from June to October: scallions, fennel, arugula, chard, kale, bok choy, and broccolini, "which I forgot to harvest," he laughed; parsnips, celery root, radishes, and carrots; and trellised Malabar spinach ("actually a tropical vine," he noted) and Mexican sour gherkins, which looked like minuscule watermelons. Jonathan and Matt spent a good five minutes scouring for the best ones for me to taste (sour, but not too sour).

"The size of your pinkie nail is a perfect one," Matt said. He pointed out the prize samples for me to pop in my mouth like Skittles. "Oh here, here! This one, this one, this one . . ."

"You missed three of them!" Jonathan said.

"We eat pounds of these things," Matt said.

Plainly, the other currency on the farm was an abiding fervor for food. Matt leaned over crimson feathers of amaranth and marveled over how it was starting to drop its seed. A heft of patience was evidently necessary for farm life as well, as a white-haired visitor interrupted. "I have real Indian pottery, and the red would be *gorgeous* in it," the visitor said, scissors in hand.

"The amaranth? You mean you want the flower?" Jonathan asked.

"Well, I don't know what it's called," she scoffed.

Matt massaged his fingers over the plant's feathery flowers and showed me the tiny seeds in his palm. They looked like uncooked quinoa. As Matt revealed, amaranth *is* related to quinoa. And, like quinoa, its seed is a gluten-free pseudograin. Amaranth is also a fast-growing plant that produces nutrient-dense food with virtually zero tending. "That's a weed. If I let these go [unharvested], I'll have every one of those germinate next spring. But how do we get that into the American diet?" Matt said.

Matt and his girlfriend, Jessica van Vlamertynghe, a full-time graphic designer, have their own farm nearby, Speckled Rooster Farm, where they grow vegetables; raise rabbits, chickens, ducks, and turkeys; and eat almost exclusively what they produce. His foodie heritage came from his Italian American family, which always cooked. "We never went out," he said. Matt's grandparents on that side were gardeners, but his great-grandparents had been farmers, and those recipes from the land were handed down. "We always ate escarole and beans or Italian wedding soup on Wednesdays—every single Wednesday," he said, explaining that escarole can survive the cold, so his great-grandparents grew it throughout the year. "I always joke that when I went to college, on the first Wednesday I was like, *Where's the soup?*"

Jonathan, on the other hand, proclaimed "Oh I'm a snacker!" when I asked what he likes to cook and eat with the vegetables they grow at Ambler. He was referring to munching on raw vegetables in the field; unlike Matt, Jonathan doesn't follow a subsistence diet because he has three kids, ages three, twelve, and eighteen, and often forgets to bring vegetables from the farm while rushing home to his family at the end of the day. "If I cook, I always start with onions and garlic . . . but if [my wife] cooks, I eat whatever she cooks because that is one of the keys to a happy life," he laughed. "[Eat] whatever your wife puts in front of you."

I asked what his relationship with food was like as a kid. "Oh, sugar, salt, and cereal," said Jonathan, who grew up in Westchester County in the eighties. "My mom still to this day can't understand why it is I ended up farming when I refused to ever help her in the garden."

We headed to the perennial area, passing broom corn, husk cherries, and winter onions. The open fields were lush with many of the plants saved from Betty Ambler's original garden—legacies from a time when the garden was also an apothecary. "There's a red bee balm that is fantastic," said Matt, then uncovered lobelia, echinacea, wood aster, and yarrow. "There's ginger in there!" added Jonathan, then showed off a row of artichokes followed by a cover crop of sorghum-sudangrass, which would be mowed to feed the soil but now looked like a field of corn.

But the pièce de résistance was a mountain of black compost that fed this fecundity, located at Ambler's edge. The source was manure from the farm's educational animals but mainly leaves, literal tons of them, collected by landscapers off thousands of lawns around town.

Jonathan didn't learn how to do any of this until his midtwenties. He caught the environmentalist bug in high school but worked first in the financial world, in sustainable investing, and then at a job in municipal recycling before coming across a farming internship in his town. "And then it took off from there," he said. "It just suited me well." His father and grandfather, on the other hand, had been suited to selling suits. Like me and many Ashkenazi Jews, Jonathan descended from garment industry ancestors. (Garment work was one of the few jobs Jews were permitted both in nineteenth-century Europe and early twentieth-century America.)

Jonathan had first farmed in Connecticut's more rural but still tony Litchfield County, where he leased land from another farmer. That farmer's farmland, in turn, was a pristine swath on the Housatonic River leased from the town, state, and federal government, which illustrates the insecure labyrinth of land access for new farmers without inherited land wealth. Finding affordable land to buy is the foremost challenge facing young farmers, according to the National Young Farmers Coalition. As such, many of them choose to rent instead: Nearly 40 percent of US farmland is leased—by and large from landowners not actively involved in agriculture. Jonathan grew mixed vegetables for a CSA, farm stand, and to be sold at the farmers' market, but "I wasn't making it," he said. "I think I grossed $12,000. And [then] I had two kids. . . . My last year I took a job at UPS over the holiday season just to try to bring in some income." He took the position at Ambler in 2012 and moved his family to Wilton.

Matt had the advantage of starting younger than Jonathan, with a degree in horticulture from the University of Connecticut, but that hadn't shielded him from other financial hurdles first-generation farmers face. "My grief rant is: Access to land wasn't my problem. It's finding markets," he said. He farmed for ten years on seventeen acres of rented land in nearby Westport, growing an acre and a half of vegetables and raising 200 pigs and 500 laying hens, before rethinking his hustle selling at farmers' markets, through an email list, and to area chefs. So he and his girlfriend bought their smaller house turned farmstead on three acres in Easton, a local food community with around twenty farms. Matt took the job at Ambler two years ago and downsized to three small wholesale accounts for eggs.

Jonathan concurred that finding land to rent wasn't that tough in Connecticut, perhaps because the state has robust programs in place to protect farmland, help transfer it to new farmers, and provide agricultural leases on state-owned land. In fact, Connecticut is one of the top twelve states in the country addressing the threat of vanishing US farmland and ranchland (18 million acres could be lost by 2040), according to the American Farmland Trust.

But there was a more insidious reason it was easy to find farmland to rent in Connecticut, a state with some of the highest property taxes in the country. "[Property owners] want the tax break," said Jonathan, glossing over a fact understood by every farmer and wealthy investor but unperceived by a general public that has romanticized the profession of farming: Farmland owners are privy not only to numerous tax exemptions and deductions but also a sizable property tax discount in most states. In Connecticut, $15,000 in farm sales or expenses wins you a property tax abatement of up to 50 percent. Not surprisingly, Jonathan had recently been approached by folks in town who wanted their land farmed, with no lease payment necessary.

So while the Van Reesemas of the world may be "super passionate" about farming, it turns out farmland—thanks to US tax policy—is a mighty convenient place to park one's wealth. Add in increasing pressure on global food production due to the Russia-Ukraine war, climate change–fueled weather events, and a population soaring to nearly ten billion people by 2050, and it's no surprise billionaires like Bill Gates, corporations, and foreign interests are quietly buying up farmland as fast as possible. They're staking their claim on a precarious food future while the rest of us—time-starved and increasingly living paycheck to paycheck—become more reliant on corporate-produced foodstuffs. Even peasants in medieval Europe had more autonomy.

"Historical returns from farmland have outpaced many other more popular investments like the S&P 500, Nasdaq, gold, and multiunit real estate," enthuses Tillable.com, an Airbnb-type start-up purporting to connect landowners and farmers, which (like Airbnb) has raised concerns about driving up the cost of land even more—in other words, ensuring a future where the superrich own the land and beginning farmers have no alternative but to rent. After all, the latter demographic isn't even likely to earn a viable living: 64 percent of small farmers have to work a second job. And then there are the farm*workers*—a human rights nightmare of largely undocumented immigrants that is the linchpin of our entire industrial agricultural system of cheap food.

Hearing Jonathan's lead about tax breaks was disheartening, as was our conversation that followed in the compost area. The sun was setting, I was shivering, and Matt had to head to Fright Night, but I made the mistake of asking Jonathan a doozy: What had to change so that small-scale farming (without the underpinning of independent wealth) could become a livable existence?

"Do you want me to be pessimistic or optimistic?" he asked, though this time I could tell Farmer Jonathan wasn't setting up a punchline. "It's not going to change. Our entire system . . . is predicated on food being inexpensive."

He explained the catch-22: If industrial civilization could keep food, a basic necessity of life, cheap, then people would continue to work their industrial jobs to pay for that cheap food, even as other necessities like land and housing skyrocketed. But keeping the production cost of food low on increasingly expensive land drove even more industrialization and technologization in farming. "The growing systems they're [building] in China and Korea, where they're huge, square blocks . . . they're superefficient, they recycle the water, they recycle the heat, they can grow tons of greens, but it's all just factory trying to work it into a different model."

(The Indian ecologist Vandana Shiva, an outspoken critic of Bill Gates, ominously calls this "digital agriculture," or farming without farmers. This is a future she believes Gates is actively shaping with his unprecedented land grab.)

"It's not human scale anymore," added Matt, encapsulating what small farmers are up against.

But as Jonathan circled back to farming within the shifting culture of Fairfield County, his tone turned more hopeful. "Now, one of the keys here for Ambler Farm that we figured out is education," he said. "Education brings in a much better margin than food."

Ambler's summer camp, for one, goes a long way toward supporting its community-funded farm. But more profound, perhaps, is the farm's apprentice program, where students in grades five through twelve, for six months out of every year, work two afternoons a week and Saturday mornings side by side with Jonathan in the production gardens. Additionally, the apprentices plant, cultivate, and harvest the educational garden's 120 handicapped-accessible raised beds. They learn to take care of the animals, rake leaves for compost, and even take on carpentry projects through an additional builders' program. There's also a Buddy Program for children with disabilities; a junior apprentice program for students in grades three through five; and My Sprout + Me for caregivers and children ages one to four.

"Culturally, people—especially around here—are very willing to pay for education," Jonathan said. "We *want* our kids to have experiences, we *want* them to experience the world." He clapped his hands on the word *want* for emphasis. Unlike Drew Duckworth's earlier professed wants, Jonathan's were imbued not just with the visceral hopes of an environmentalist and farmer but those of a parent, that perhaps in the next generation, things could truly change. "And so education, to pair it with farming, is phenomenal. It's a very easy pair."

Ashley summed up the success of their educational programs with another angle: "[The kids] are not on their phones, number one."

Both of Ashley's sons, now twenty-two and twenty, had gone through the apprenticeship in its early days, not long after being launched "with just a handful of kids" by Kevin Meehan, a former elementary school science teacher who lived in the farm's historic 1800s Yellow House. The program now enrolls 130 students, a significant number in a school district with about 2,500 kids in grades five through twelve.

"My son will be a senior in college, and he along with three or four of his friends from the apprentice program have all decided that they want to go into environmental studies [or] sustainable agriculture or that type of field," she said. "The program has really shaped what their path is."

"Have any of the apprentices gone into *farming*?" I asked.

"I can't think of anybody," she said, before heading to Fright Night.

"If they're smart, no," Jonathan laughed. "But they're gonna love farms!"

Still, he seemed hopeful. "I would think that in the long run, if you want people to live differently, you have to *show* them how to live differently, but you can't do it in a way that is combative or confrontational," he said. "They just have to experience it. So the more kids you can bring through here and [have them] eating foods and just seeing—*Oh look! This grows!*—then maybe they'll want to plant something or have something growing. And that will be that slow spark that will eventually lead to change," he said.

Yet the question that lingered was whether we have time for gradual change. This is an imperiled and uncertain age, after all. But if there's a model for a way forward—if there's a David act against Goliath greed—a community investing in its own land, in its own local food, and in the education of its own generation of food activists (if not yet farmers) sure seems like the closest thing we have to an answer.

Just as Matt hopped on the shiny red tractor to head to the barn and the sun plunked below the horizon, I realized our last-minute interview had turned into two and a half hours. ("One of the great things about our business is that you can do mindless work and then you can think. . . . It leads to a lot of philosophers," Matt had laughed earlier.)

Still, he couldn't help squeezing in a tagline. "I think the real key to ending this [conversation] is neither of us would at any point change what we do. And I think that's the thing—we're just so committed. It doesn't work financially, it doesn't work on an energy level..."

"It doesn't always work for your families!" interjected Jonathan.

"But we wouldn't change what we do," Matt said before driving away.

"I don't know what it's gonna look like in ten years," Jonathan mused on the future as we walked toward the carriage house. Kids were streaming up the hill toward the barn. A small boy toting a pumpkin bucket just missed me as he sprinted to join his friends. "I think we work on one apprentice at a time."

FOUR

Round Table Farm

"We're better off for all that we let in."

—Emily Saliers

THE TREES WERE BARE AND THE GROUND BRUSHED WITH FROST by the time I hit the final eight miles of state highway leading to the center of Massachusetts. But save for the flora, I could have been anywhere in rural America. Off the interstate exit, my first landmarks were an opposing McDonald's and a Wendy's. A LET'S GO BRANDON banner flashed by, then a roadside stand bearing a blue and white cow statue and a delicious-looking remnant of the area's agricultural history: HOMEMADE ICE CREAM–RONDEAU'S DAIRY BAR. After an underpass beneath a rusted bridge, discount beer and wine, a cemetery of slanting headstones, and a Gothic church advertising a food pantry, I spotted the historical placard: HARDWICK 1739.

In the town of Hardwick's genealogical registry, townsfolk with the name Robinson appear all the way back to that founding date. And the Robinson family traces to the *Mayflower* and England before. For four generations, descendants of those forbearers farmed at the Robinson Farm, which I had traveled that day to visit. Ray Robinson, now seventy, is one of them. Ray's great-grandfather purchased the farm and started dairying in 1892, and Ray did the same, he and his wife, Pam, raising their "Brady Bunch of six" children there (three and three from prior marriages; the couple has been married thirty-four years).

Except this isn't a story about Ray or the Robinson family—although they play a laudable part in it—because the Robinson Farm in name is no longer. Ray and Pam's children didn't want to take over the family farm. And so I had come that Saturday morning to meet and write about Marlo Stein and Archer Meier, the young queer farmers turned cheesemakers who moved to Hardwick and started a cheese and flower farm on the site of the historic Robinson Farm in 2021.

Theirs is a scenario playing out all across the United States: More than a third of America's 3.4 million farmers are now over the age of sixty-five. And as this bastion retires, they—like the Robinson family—will face painful choices about whether to sell the family farm and, if so, how to preserve the land and their legacy.

The name Marlo and Archer chose for their new endeavor, Round Table Farm, might be viewed as a metaphor for our culture's chance for renewal. As the two of them were brave enough to show me and their new community, the face of rural America is changing. But if small farming communities like Hardwick are to survive, our expectations for earning a livelihood off the land might have to change too.

"I DIDN'T KNOW THAT farming was a profession I could choose," said twenty-eight-year-old Marlo, who grew up in the Boston suburb of Newton, Massachusetts, where her mother was an architect turned stay-at-home mom and her dad worked in software engineering. "I thought it was something that was just passed down in families."[1]

---

1 Marlo wasn't wrong. Only a quarter of the 93 million acres of US farmland transferred between 2015 and 2019 was sold between nonrelatives; although, as discussed in the previous chapter, developers and foreign corporations are increasingly becoming buyers. In March 2023, Senator Josh Hawley of Missouri introduced legislation to prohibit Chinese entities from purchasing American farmland. Chinese ownership of American farmland increased 30 percent between 2019 and 2020 to make up an estimated total of 384,000 acres, although other countries actually own more than that.

Marlo guessed she is three or four generations removed from her family's farming history, although her mom and grandparents, of Swedish ancestry, grew up in Hardwick's agricultural Worcester County. Her dad's side of the family is Jewish, and her grandmother's grandfather had been a butcher in Vermont after resettling in America. "Not having family land . . . it felt really meaningful to get to take on a farm from a family that didn't have another generation to take it on," she said.

Marlo and Archer had just welcomed me into the 1700s farmhouse after touring me around the barns, former milking parlor, and creamery, set on thirty-eight of the original farm's more than 200 acres. Sunlight streamed in from the windows gazing at the apple orchard out back, which had been planted by Ray Robinson's father, Raymond G. Robinson Sr.

Marlo grew up in a home with fruit trees in the backyard and a mom who made scratch applesauce and hosted jam-making parties. "I learned to sew and we had a garden," she said. "It was funny—I was very in touch with a lot of homestead skills growing up in suburban Newton."

Archer, twenty-seven, pointed out the greenhouse and quarter-acre flower field, now brown but which that summer had been lush with a rainbow of dahlias. "We've had a good hard frost . . . but you can still see the ghosts of what was," he smirked.

We walked into their lime-colored kitchen anchored by a (yes, round) wooden table. The fridge was bedecked with Polaroids and a magnet that said LIFE'S A GARDEN. DIG IT! We sat and Archer picked up their cat, Velvet, for a snuggle. Marlo continued her story about her path from suburbia to farming.

In high school, Marlo spent four months at Maine Coast Semester, an experiential education program that centered around a farm on campus. "That was the first time that I met people who had *chosen* to become farmers," she said. "[I] saw the intersections of social justice, food justice, the work that I wanted to be doing outside with my body. There was a tangible impact to it. I never went back to anything else." Later, she apprenticed at women-owned Merrifield Farm in Maine and worked as a camp counselor for Farm & Wilderness camps in Vermont over summer breaks from Massachusetts's Smith College, where she and Archer met on the rugby team. Marlo was the sophomore in charge of welcoming freshmen rookies, which included Archer.

Archer *mm-hmm*'d knowingly, but divulged, "I really don't think of myself as a career farmer." Unlike Marlo, Archer has pursued a parallel profession. At Smith, he studied sociology and education and learned sign language, and when we met was teaching off the farm at a school for the deaf. He recently started nursing school.

Part-time farming was a way of life Archer actually had a vision of as a seven-year-old, middle-class kid in urban Minneapolis. "I didn't see myself as a farmer, like as my job, but that I lived on a farm and I had animals and I worked with those animals every day," he said. This foretelling wasn't wholly unexpected: Archer's great-grandparents had farmed in Minnesota after also emigrating from Sweden ("a funny connection," Marlo noted), and his grandmother grew up on that farm. Archer also loved horses and went to YMCA horse camp.

"And then, as I started realizing that I was queer in my early teens and onward, I was like, *Oh, I can't live on a farm. I will never live on a farm because I'm queer.* So I'll live in the city because that's where queer people live. That's where community is available."

But after graduating from Smith, Archer joined Marlo in the rural Midwest, where the two worked on organic farms in Wisconsin and Minnesota. "It took a lot of coming back to that idea and opening up and saying, *You know what? Queer people live all over the place,*" Archer said. "I believe that more and more queer and trans people will be moving rurally in the future. It's happening now. I feel very confident that . . . we will in the next few years not be the only queer couple that lives in town."

Archer said this matter-of-factly, without a hint of activism. But I couldn't help but feel the reverberation of that statement in parts of the country that might be less than welcoming to a queer influx. He admitted this was "less scary" to say in a more liberal state like Massachusetts, although central Massachusetts, where Round Table Farm sits, is decisively more conservative. (Hardwick was split fifty-fifty in the 2020 presidential election.)

"We lived pretty rurally [in Minnesota]. I was a lot more nervous," Marlo said.

"Yeah," Archer said.

But when I asked what the two most wanted me to convey about their story as self-professed queer farmers, Archer didn't hesitate. "I think that it's a disservice to queer people to have the idea that we can only live in cities, and that cities are the only place that we'll be accepted and can be safe." He pointed out that, generally speaking, cities are increasingly *un*safe.

Then, too, there's the therapeutic benefit outdoor physical labor and connection to the land offers queer farmers, Archer added. "To . . . just feel that closeness to your world is a very queer-affirming thing. And for me, specifically, a trans-affirming thing as well."

Because Archer came to farming as a lifestyle, with an outside career, their farm work could be guided not by financial or societal pressures but instead "a sense of being in my fullness as a person." There was the childhood dream realized of working closely with animals. At Round Table, Archer explained, there's *a lot of space.*

"I can feel very free to just be myself in this space in a way that I never felt living in the city, even among a lot of queer community," he said. "And here, we don't really have any queer community that's close to us. But I still feel such a sense of spaciousness that I never got to feel. I really want other people to be able to experience that joy."

NOW UNDOUBTEDLY, ARCHER AND Marlo are in the honeymoon phase of farming. As I write this, 1,500 tulips are sprouting in their hoop house for Round Table's second season. And, by the time you read this, the two will have had their literal honeymoon: After I asked about a Post-it marked WEDDING CHZ in the creamery's final aging room, Marlo giggled and told me they were getting married the following summer. The tagged wheels of cheese were crafted the week they got engaged and would be served at their wedding.

Still, the joy Archer had described was palpable enough to feel enduring. After he gave their new cow some behind-the-ear scratches in the pasture, the couple walked me to the smaller barn to meet their new goat herd—a mix of Nubian, Saanen, and Alpine breeds—that would be used to clear invasive brush and provide milk for goat cheese.

"Are you comfortable going in with the goats?" Marlo asked as I tracked Archer's Blundstone boots heading into the pen. I paused, wondering if the herd was aggressive, then watched them charge Archer for nuzzling.

"You're the sweetheart. You're the cuddlebug," Archer said, kissing the one with the big horns.

Later, Marlo and Archer cracked me up as they outfitted me in a hairnet and disinfected Crocs to enter Round Table's inner sanctum: the "make" room for their aged raw milk cheeses. I looked like a giant nerd, and they looked the epitome of Gen Z farmer cool: Archer wore a mesh-panel cap, Patagonia pullover, tortoiseshell glasses, and an Apple watch; Marlo, who escorted me into the room, was fresh-faced even in a hairnet, her cheeks matching her rose-colored corduroy jacket. Both of them had nose rings. Archer assured me the couple often hit up Zoom meetings mid-cheesemaking in their cheese coats.

"My therapist loves when I do therapy in [the hairnet]," Marlo chuckled.

But it was after eyeing the shiny cheese vat the size of a hot tub and asking about the working vibe in the (actually) sterile fluorescent-lit room that I got a full picture of the pair's joie de vivre. I was expecting them to toss off what kind of music they liked but instead got a rundown of an entire audio-curated cheesemaking day.

"We listen to a lot of news in the morning . . . that's like the morning mood," Marlo said, describing their start in the creamery around 7 a.m. After cleaning, the pair split up, with Archer staying to prep while Marlo hitched up the mini trailer tank for the forty-minute drive to pick up milk from a local farm. ("Local" is a relative term, since two-thirds of the state's dairy farms have shut down in the past twenty-five years.)

"You usually have your smoothie or your coffee or your bowl," Archer said, grinning at Marlo. "You listen to your podcasts . . ."

"I listen to *We Can Do Hard Things*," Marlo said. "I think of it as my morning conversations."

But later, after four hours of stirring, coagulating, curding, draining, packing, and pressing, that's when the music really got going. "The cleanup process after making cheese is pretty intensive, and then we're always listening to music—a lot of folk music," Marlo said, citing Brandi Carlile and Indigo Girls as sing-along faves. "I'll listen to a lot of pop music if I just need to get a little more pumped up."

"Cruising on the cleaning!" Archer chimed in.

"The cleaning is always the final push that we need some good energy for," Marlo laughed.

Pop-y playlists? Therapists? Smoothies? I'm a Vitamix fan, and tending mental health is critical these days—especially for LGBTQ adults and farmers, who both suffer markedly higher rates of depression and suicide than the general population. But I took a wager this was a peppier approach than that of the farm's prior generations, relentlessly rising with the roosters to go milk the cows. Dairy farming, after all, is a notorious grind. That reality, coupled with perpetually low milk prices due to US government subsidies, growing competition from plant-based milks, and pressure from developers to sell, led 10,000 dairy farms in New England to shutter over the past half century.

"Once we start milking, our schedule will change a little bit," Marlo said with a nervous laugh when I asked about the plan for the goats. As for the cows, Marlo and Archer only kept two from the Robinsons' original herd of around forty. Instead, they're procuring raw milk for their cheeses from other area farms to support those endangered businesses and give themselves some breathing room while they get Round Table Farm up and running. "But we really believe the farm should fit us and our lifestyle, as opposed to the other way around. And so we won't be 5 a.m. milkers. We'll probably milk our animals around 7."

I admired their intentionality, but it was here that I couldn't help envisioning a smash cut to five years in the future, milk pails sloshing as the two of them wailed "Closer to Fine" in the midst of a Massachusetts blizzard.

Was work-life balance as a new farmer actually achievable? Benji and Carys at Mahonia Gardens in Oregon had found it, but it had taken *years*. And if so, would more young Americans—perhaps less willing to suffer the sacrifices of generations past—be willing to become farmers?

We peeked in on another aging room, this one with racks of fresh loaves of a "bloomy and bright" Taleggio style called Sweet Pea, which would be aged for sixty days before being sold. (Square cheeses are called loaves; the round ones are wheels.) This was one of three of the former farm's award-winning cheeses Ray and Pam Robinson taught the young couple to make. The Robinsons had been relatively new to cheesemaking themselves. The financial benefit—and potential reduced workload—for farmers offering "value-added" products like cheese was one of the reasons they installed the creamery in 2011. In fact, for decades the Robinson Farm had been a conventional dairy shipping wholesale milk. It was Ray's daughter, Gina, who set the farm on a path to a profitable niche market after becoming interested in organic farming at Oberlin College in Ohio and starting a raw milk side hustle with four cows on the family farm. That success led the family to convert to grass-fed practices and become certified organic. Behold, the power of the purse: These changes were what their customers asked for.

"[The Robinsons] knew that they wanted to create a farm that was financially feasible for young farmers to come into," Marlo said. "They knew large dairy wasn't an option. No one our age would be able to come in and buy a medium dairy—large in our area—and be able to make a living off of it."

But whereas the Robinsons had forayed into cheese as a bonus to their exhaustive farmstead operation—milking their own closed herd of cows; grazing them on their own land; harvesting their own hay for winter feed; making their own cheese; and selling their own beef, eggs, raw milk, and seasonal vegetables, all while maintaining costly organic certification—Marlo and Archer are taking a potentially more flexible and profitable approach.

In addition to selling small-batch cheese and flowers, they plan to turn the property's farmhand house into an Airbnb. They're also contemplating becoming an event space for weddings, given the flower farm and picturesque barn. Eventually, they want to bolster their cow herd and sell raw milk, which had drawn customers all the way from Boston in the Robinsons' day.

The vision doesn't stop there. "We're hoping to buy a pasteurizer because we'd really love to do cream cheese," Archer told me in the farm store adjacent to the creamery, which they plan to stock with local goodies like maple syrup from Hardwick Sugar Shack. They've already created a Queer Cheese Subscription. And later that afternoon, they were hosting a beer and cheese pairing in town with a microbrewery. It seems there is no end to their creativity.

But there is also a meticulously crafted business plan, a necessity when the couple came to farm ownership earlier than expected. After farming in the Midwest, Archer and Marlo initially moved back to Smith's college town of Northampton, excited to take a break from rural life and enjoy the city's lively LGBTQ community. Then came COVID, and they instead found themselves isolating with Marlo's mom, who lived in town.

"It was amazing to get time with our family, but we gave up on this midtwenties, fun social life," Marlo said.

"We also realized that we didn't actually *need* it," Archer said. "It was kind of an idea that we thought we were *supposed* to do, being twenty-four and twenty-five."

Not long after, Marlo came across the listing for Robinson Farm. "It had all the things that I would be looking for in a farm," she said. But at its original price, at which it had been listed for two years, it was just a dream. "And then I got a notification that the price had dropped . . . by well over half." The Robinsons had reparceled the property.

So in October 2020, at the height of the pandemic, Marlo and Archer and Marlo's mom drove out to Hardwick to look at the farm.

"And all of us walked away being like, I think we could do it," Marlo said.

"It was like, *Oh yeah, no, this is perfect,*" Archer said. "We were walking around, just, *Wow!*, mind-blown with how much potential there was."

Archer and Marlo then realized they could bring the social life to *them*, although with a loftier mission than their original midtwenties partying plans. "We had so many friends and family [quarantining] in cities. We were like, *I just wish that we could bring you out somewhere*," Marlo said. "It was a huge, huge, push for us to take on this place that is really massive and feels available to be filled with people," Archer added.

Since then, the pair has had an open-door policy: Close friends have stayed for six months. An acquaintance crashed for two weeks. "And so a big piece of our ethic here, too, is *How can we gather?*" Marlo said. Ultimately, the couple hopes to foster more formal community space, such as a shared commercial kitchen for local food businesses.

These are heady ideals that they readily admit are contemplable "because we do have access to family wealth and support," Marlo said. Still, neither comes from the upper crust of chapters past. They have a mortgage. Archer essentially works two jobs.

When I called up Pam and Ray Robinson a few months later, Pam summarized the young farmers' circumstances in straightforward New England farmerspeak: "You're not gonna get rich doing it. So it has to be a lifestyle choice, really."

I asked why none of their six children had picked that way of life.

They said Gina, the daughter who influenced them to switch to regenerative farming practices, had been their initial hope to take over the farm. "But then she met a man on the internet and moved to Indiana," Pam sighed. "Most of [our children] are in the cities. A lot of them are far away. They're all over the country."

"There's a lot of good reasons not to want to be in farming," Ray said.

"They wanted to make more money," Pam said.

"Yeah, it's . . ." Ray paused. "It's a very hard thing to do. It's very hard."

When I called the Robinsons, it was close to 4 p.m. in Vermont, where they moved to ski and be closer to their grandkids. Pam said they were tired. They had spent the day at the "little farming job" she and Ray took on two days a week to supplement for groceries, given the soaring cost of food and the economy's recent impact on their retirement account.

The two also sounded understandably emotional. "We're still grieving in a way. It's hard to go back [to visit]. We have a lot of friends we left and, you know, generations of family. Well, there really isn't much family there left, just history," she said.

"And familiarity, you know?" Ray added. "Whenever I do go back, it feels like it's still my barn."

"Well, he still has a tractor in the barn," Pam quipped. "Because we still have 120 acres there that we haven't sold yet."

While the Robinsons initially sold Archer and Marlo the farm with thirty-eight acres and the option to buy an additional seventy, they planned to sell the remainder as ten-acre "farmettes" in the hopes of attracting new homesteader-types to Hardwick. The Robinsons also donated thirty-three of the farm's acres to the East Quabbin Land Trust. Land trusts are a community model of land ownership increasing in popularity as a means to preserve farmland for sustainable agriculture. "It was important to us to have some part of the land forever green," Pam said. Another incentive: "It helped with our capital gains," she revealed. "We plan to put our ashes there; that's part of the deal. So when we die we're going to the night pasture." Pam said this deadpan; Ray was laughing.

Thus will be a morbid though poignant ending—nay, regeneration—of a quintessentially twenty-first-century American story about the descendants of a first family of Massachusetts passing the torch to the queer descendants of Swedish and Jewish immigrants. "We're just really happy that it stayed a farm, and that was our goal," Pam said. "That makes us feel better about having to walk away."

And what will be Archer and Marlo's biggest challenge for making the farm successful?

"I think trying to balance their time between the different phases of the operation," Ray said diplomatically.

"They're young, and they have a lot of time, and they have money so that they don't have to jump into everything all at once," Pam said. "They're smart. They're very, very smart, quick learners."

⁂

BUT IN THE END, it may be Marlo and Archer's commitment to community that allows them to prosper—albeit by a more holistic metric than our corporatocracy currently measures. This was a lesson they learned working on farms in the rural exurbs of Minneapolis, where "the biggest alienating thing" was "city folks coming in just to buy property."

Recently, the new farmers in Hardwick had a community cause for which to rally: a fight over the future of Great Meadowbrook Farm, a 360-acre historic property down the road from Round Table Farm put up for sale. Initially, it had been snatched up for $2.6 million by entrepreneurs from the city of Worcester, Massachusetts, who tried to install a cannabis enterprise before the town shut down the deal. Then, other developers tried to buy the property to foist a thoroughbred race track and gambling facility on the bucolic farm town.

The farmland was ostensibly safeguarded by Massachusetts's Agricultural Preservation Restriction (APR) program, which has protected more than 800 farms and 68,000 acres in the state since 1979, but the developers made the loophole argument that race horses could be considered an agricultural product. Ray Robinson had known the late owner of Great Meadowbrook, whose farm had also been in the family for generations. "Ray was like, 'Chet Goodfield would be rolling over in his grave.... He wanted this to be a farm forever,'" Marlo said. Canvassing and campaigning by Hardwick's citizens ensued. The week I was there, the board of selectmen voted to turn the racetrack plan down.

In January 2023, the town's residents rejected the plan in a referendum, 830–312. A proposal for the developers to resubmit is still pending. But even if it's overturned, I suspect it won't be the last time the small farming town has to fend off developers. Marlo and Archer plan to stay politically involved in town.

"There aren't a lot of young people [in this town], and if we want to see the future for our kids that we would hope for [here], then we gotta be the ones saying what we want," Marlo said. A few minutes later, she raved about the pastries at a start-up organic bakery she suggested I hit up on my way out of town. The owners of that new venture had come from California; they had sold a big bakery business that supplied supermarkets like Safeway and Trader Joe's, and decided to retire with an encore career in Hardwick.

"They were welcomed from *California*?" I asked.

"There was some contention," Archer said.

"We'll see if they stick around," Marlo added.

Archer and Marlo, on the other hand, said their own transition to the 2,667-person town of Hardwick, thus far, has been easier than expected—despite being obvious newcomers. ("If we were queer farmers of color, I don't necessarily think this conversation would be the same," Archer noted. Hardwick is 87 percent white.)

In a town that never lost its agrarian tradition, "being a farmer gives me a specific 'in' with anybody," Archer said.[2] "I can always talk to people about the weather and how things are growing. . . . I think a lot of lack of [queer] acceptance comes from, *I don't understand*

---

2 The reason Hardwick has stayed a farming community, the Robinsons explained, is the 38-square-mile Quabbin Reservoir, which was constructed by sinking four neighboring towns in the 1930s. One of the largest unfiltered municipal water sources in America, it supplies greater Boston with drinking water and has kept central Massachusetts rural: No major highways can be constructed through the center of the state.

who you are; I don't understand what you're doing. Being farmers, it's just a general *I understand what you're doing here.*"

And in a world where agricultural land is being lost to residential, commercial, and industrial development at an astonishing rate, there's something far more important in rural places than one's gender identity, Marlo said: "showing up for each other and showing up for the community."

FIVE

Assawaga Farm

"After all that men could do had failed, the
Martians were destroyed and humanity was
saved by the littlest things, which God, in
His wisdom, had put upon this Earth."

—Barré Lyndon, *The War of the Worlds (1953)*

WHEN YOKO TAKEMURA AND HER HUSBAND, ALEX CARPENTER, set out to start a farm of their own, they considered leaving the United States for Japan. After all, there was the draw of culture: Yoko's parents are Japanese, and she spent a good portion of her own childhood, including college, in Tokyo. Then there was the lure of food: The couple wanted to grow Japanese vegetables— varieties like the sweet, slightly spicy, green *komatsuna*, *shishito* peppers, and *gobo* (burdock root) that Yoko ate during her youth. Locally farmed Japanese produce isn't easy to find stateside, so they had first imagined finding a niche growing it for New York City's Asian and Asian cuisine–appreciating populations. Recalled Alex: "We were looking for land all through [New York's] Hudson Valley, western Connecticut, even up into southwestern Vermont and [Massachusetts's] the Berkshires, but there's so much competition; the land prices are so high."

In addition to high prices, finding suitable soil in the Northeast was a big issue for Alex and Yoko: They cross-referenced every listing with the USDA Natural Resources Conservation Service Web Soil Survey database (websoilsurvey.nrcs.usda.gov), which has soil maps and information for more than 95 percent of counties in the United States. "Immediately, you would be like, *Oh, that's why that land is for sale*," Alex said. "There's no access to water. . . . It's nothing but rock. It's 3 pH. [A pH range of 6 to 7 is ideal for most plants.] It's underneath powerlines. It has a gas pipeline running through it. So they don't disclose that [last one]. But you have to do your due diligence."

The two initially searched for land for a year and found nothing. That's when they couldn't ignore the biggest pull to Japan for new farmers: "Land is basically free," Yoko said. "You can't, you don't really own it. But the young people, they've all fled the countryside, so people just have this land, and they want people to do something with it."

So Yoko and Alex went to Japan to scout and talk firsthand with farmers. It turned out the available farmland, while indeed free, didn't exist in the wide, open swaths we take for granted here in America. In Japan, "It's just fragments everywhere," Yoko explained while culling potatoes the August afternoon we first spoke by phone. "A lot of our [Japanese] friends who farm there tend to acquire more and more little parcels, but that one's twenty minutes away, that one's thirty minutes away—they're just driving between all these little plots."

Yoko has a master's in environmental sustainability from New York's Columbia University, and Alex has a degree in human ecology from Maine's College of the Atlantic. Unquestionably, they wanted to farm organically—a difficult mission, they soon realized, in a country like Japan with conventionally farmed rice paddies everywhere. (The draining water, laden with commercial herbicides and pesticides, would inevitably contaminate their plots.) Finally, there was the question of how they would earn a living. Outsiders often think food in Japan sells at a premium, à la Kobe beef and rarefied melons. Yet day-to-day fare is a different story. "Vegetables [in Japan] are so cheap—*dirt* cheap," Alex said.

As in the United States, agriculture in Japan is heavily subsidized. But unlike the United States, which focuses its crop subsidies on commodity corn and soy to feed factory-farmed livestock, make highly processed food, and produce biofuels, Japan directly subsidizes—gasp!—vegetables. This is fantastic for the waistlines of the Japanese, who have a 4.5 percent obesity rate (versus 42 percent in the United States), yet it is a factor hindering the income of Japan's small vegetable farmers, the majority of whom grow produce as a side job.

"And every individual vegetable has to be wrapped in plastic," Yoko laughed.

"Also, I mean, I'm a white guy. I don't speak the language. Even if I spoke the language, I'd still always be an outsider," quipped Alex, who has blue eyes, a red and gray beard, and stands a head taller than Yoko. (She elaborated: "Most farmers don't speak any English. Most people in Japan don't really speak English.") A few minutes earlier, Alex's voice sounded tinged with regret. "We could have had a good life there. But it would have been hard."

"It would have been really hard," Yoko affirmed.

IT WAS MIDAFTERNOON, AND I had driven forty miles on all back roads from Hardwick, Massachusetts, to reach Assawaga Farm, a one-acre, no-till, beyond-organic vegetable farm. Yoko, thirty-six, and Alex, forty, started Assawaga Farm in 2016 in Putnam, in the northeastern notch of Connecticut. Alex grew up in the western part of the state, in Brookfield, but this was an area he had never visited as a kid (nor I; such was the pre-Internet childhood). But a few weeks after returning from Japan, he and Yoko decided to carry on their land search in the United States and checked realtor.com. "This was on there, and it was like, *That looks pretty good. And the price is pretty good*," Alex recalled.

"This" was twenty-two acres of vacant land cloistered by woods and wetlands, bordered on the west by the Five Mile River—a private landscape unimaginable in densely populated Japan. "It's like our own little nature preserve out there," Alex said. (The river was called *assagawa*, or "place between" by the native Nipmuc, the Algonquian-speaking people who originally stewarded the land of northeastern Connecticut, northern Rhode Island, and central Massachusetts.) What's more, it was only an hour's drive to Boston, where Alex and Yoko now sell vegetables at the Brookline Farmers' Market.

Alex and Yoko bought the land for $117,000 with some savings from earlier careers, or what Yoko calls her "long-winded path to farming": After graduating from Japan's Waseda University, she worked at Deutsche Bank in Tokyo, Manila, and Singapore before heading to grad school at Columbia. Then, while doing a stint at a management consulting company ("I was on the F-1 student visa, so I had to look for legitimate work"), she joined the CSA for Wind-flower Farm in Valley Falls, New York. One serendipitous day, she took a trip to the farm's visiting day. "On the ride back home to New York City, I was like, *How do I get myself on a farm?* I couldn't get it out of my head," she said. She landed her first farming job at River-bank Farm in Roxbury, Connecticut, shortly after.

Yoko was already dating Alex at the time; they had met at the community garden in the Bed-Stuy neighborhood of Brooklyn, where she lived. Alex never farmed before Assawaga; he spent fifteen years building pedalboards for Analog Man in Bethel, Connecticut—widely considered one of the best guitar pedal companies in the world, my guitar fanatic husband attests. (Alex still builds pedals in their winter offseason; Yoko wires circuit boards.)

"The western side of the state, it holds a special place in my heart ... [but] we never would have been able to afford a farm over there," Alex said. "There's a lot of wealth and not a lot of opportunity on that side of the state. And on this side of the state, there's not a lot of wealth. There's a lot of opportunity, but there's also a lot of ... Republicans," his eyes twinkled as the three of us burst into laughter.

"Really, I didn't notice any of the signs on my drive here," I deadpanned.

Whereas Yoko was sunny and lively, Alex was soft-spoken and considered, with a touch of curmudgeon. He said he spent much of his teenage years in the woods—there was a trail from his house to the local high school—and his quiet presence was that of someone who had never left nature. When I asked Alex why he farmed, he didn't hesitate. "Just being outside and being free. . . . I enjoy going to the market, but it's not . . . I could do without that. If I could just be outside growing stuff, I'd be happy."

Unlike many other farmers I had met so far, who talked a lot about finding community, I got the feeling Alex and Yoko appreciate their seclusion.[1] Actually, it wasn't just a hunch: I missed the entrance to the farm twice because it is intentionally unmarked. "We put a sign out during farm stand hours. We like to be hidden," Yoko said. And it was nearly an hour into the interview when Alex finally emerged from the barn. "I decided to stop being antisocial," he said with a warming laugh.

When the couple bought the land that would become Assawaga Farm, it was rich with life—"prime agricultural soil," Alex said. But it was raw. "There was nothing here," attested Yoko. They meticulously crafted the property from the ground up, putting in a driveway, drilling a well, and building their greenhouse and eco-modern barn/residence almost entirely themselves, with professionals putting in the big timbers and the two of them filling in the walls, doors, windows, and floors (in winter, no less). Neither Alex nor Yoko has a carpentry background. "Alex watched a lot of YouTube videos," Yoko said.

Then, to heat their propagation area in the greenhouse, Alex built

---

1 If you, too, are an introvert, I think it's important to know: Yes, humans evolved for social connectivity; yes, many ancestral societies farmed communally, and that model has tremendous potential in the modern agricultural world. But if you're drawn to the happy hermit path of regenerative farming, Alex and Yoko are a great example of how to be successful.

a rocket stove, which they used until purchasing a real heater the spring after I arrived. "We were just living really frugally . . . and so now we're in a great place financially," she said. They were finally able to hire their first employees the previous season, although it had been hard to hold on to them, given the lack of housing in the area and high gas prices. Their part-timer quit in early August and their full-timer quit the week before I arrived, both with no notice, Yoko said, adding woefully: "Young people don't want confrontation. It's just easier to ghost people, I think." Thankfully, though, they are now in a position to purchase employee housing in town. "If we knew that we would be somewhat successful, maybe we would have borrowed far more money. But I'm happy the way we did it. In Japan, it's not that common to be in debt."

Yoko was wearing a maroon thermal and drawstring cargo pants, and put on a straw gardening hat to walk me around the production area of the farm, formerly a hayfield that now encompassed eight-tenths of an acre. "We have just by sheer chance 100 hundred-foot beds. It came out exact," she said, her dimples belying that statement. For Yoko and Alex, I could see perfection was embodied rather than happenstance.

In a farming area the size of many people's backyards, they grow more than fifty types of vegetables, including over 150 varieties, over the course of a season. That Saturday morning, Yoko had been harvesting for their Sunday (now Saturday) farm stand in the barn, which would include Hakurei turnips, Black Futsu squash, Nabechan scallions, Daikon radishes, and the three popular Asian greens they grow year-round—komatsuna, mizuna, and bok choy—impeccably arranged next to laminated signs with Yoko's graceful illustration of each vegetable, the name scribed in English and Japanese, the price, and preparation suggestions. DON'T LET ITS LOOKS FOOL YOU! asserted one with the likeness of a hairy-rooted celeriac. GREAT IN SOUPS, SLAWS, GRATINS, MASHED, AND MORE!

I HAD FIRST BEEN drawn to Yoko and Alex's story because of my admiration for Japanese culture and food, especially vegetables. I could eat those slender violet eggplants or Japanese sweet potatoes—*satsumaimo*—every day of my life.

So I initially thought this would be a chapter about cultural reclamation through farming, especially after learning about Yoko's uprooted childhood: She was born in Venezuela and lived in Japan betwixt some primary school years in the United States and high school in Australia. After all, I had read about other young Asian Americans turning to farming as a way of reconnecting to their heritage—like Kristyn Leach, a noted Korean American farmer in Northern California who's creating an Asian heirloom seed bank, or Mai Nguyen (also in California), a Vietnamese American climate scientist turned heritage grain farmer who cofounded the Asian American Farmers Alliance.

And without a doubt, Yoko is reconnecting to her native culture by growing and sharing the Japanese vegetables she loves, like fresh edamame. Assawaga Farm also supports Asian seed keepers, such as Kitazawa Seed Company, which survived the Kitazawa family's internment during World War II and sells seeds for more than 500 Asian heirloom varieties. While Alex may not be of Japanese origin, cultural reclamation certainly struck a chord with him when I explained it was a theme of this book (along with reconnecting to nature, food, and community in a modern world). "You mean doing something that *means* something?" he laughed, besting my elevator pitch.

Yet, the culture the two seemed most invested in reclaiming was in the earth beneath their feet: "What's really awesome about this year—every year—is our soil is improving. And we can tell just by how things grow," Yoko gushed before showing me where they make their biologically active compost.

⁂

FROM THE BEGINNING, YOKO and Alex decided to make Assawaga a no-till farm. They had seen the system in practice at nearby Tobacco Road Farm, stewarded by renowned organic farmer and author Bryan O'Hara, whom they consider a mentor. "There were no weeds. Everything grows gorgeously," Alex said.

No-till, by its simplest definition, means not digging up or turning the soil—a timeworn technique used to ready the soil for planting. But whereas early horticultural societies may have used sticks and hoes to till, the practice became more disturbing (both geologically and philosophically) over the ages—with ox-drawn plows emerging in Eurasia around 4000 BC and, later, the invention of the "self-scouring" steel plow in the early nineteenth century by an American blacksmith named John Deere. (It was John Froelich who came up with the gas-powered tractor just over a half century later.)

The last century-plus of tilling and plowing the planet with monstrous machines—not to mention intensive livestock production and spilling pesticides into the land—has degraded a third of the earth's topsoil, according to the Food and Agriculture Organization of the United Nations, and sent catastrophic levels of carbon into the atmosphere.[2] We need this now vanishing uppermost layer of the earth's surface to grow 95 percent of our food, and the news isn't good: A recent study calculated that more than 90 percent of

---

2 Technically, plowing is a type of tilling. Tilling refers to agitating the soil to ready it for planting, while plowing is a deeper dig with the goal of turning over large masses of soil. And while we're clarifying definitions: Soil *degradation* refers to a diminishment in soil quality and its ability to support life. Soil *erosion* is when soil is literally washed or blown away. Erosion is one type of degradation.

conventional farmland is eroding, with 16 percent facing a life span of less than a century.

And then there is tilling's incalculable impact on the soil microbiome, with its complex interactions between bacteria, fungi, protists, nematodes, and other microorganisms—not to mention larger creatures, such as the mighty earthworm, nature's tiller, capable of breaking down rock. As with the human microbiome—the trillions of microorganisms that live on and inside our bodies, with which we have evolved and which are pivotal to our health—this ecosystem is a tiny universe that science, let alone the agricultural industry, is only beginning to understand.

Of course, the soil microbiome and human microbiome are inextricably linked. As we've increasingly doused our crops with pesticides and herbicides, the eons-tuned balance of microbiota has disappeared from our guts and even our lungs, contributing to the rise—the latest science is revealing—of autoimmune diseases, including ulcerative colitis and lupus; autism; asthma; and cancer. The use of the herbicide glyphosate, the main ingredient in Bayer's (formerly Monsanto's) Roundup, for instance, increased nearly 2,000 percent—yes, you read that right—from 1992 to 2016.

What's more is the nutritional loss that's transpired as we've destroyed our soils. One of the pivotal roles microbes play in the soil is cycling nutrients. One study comparing USDA nutrient data for forty-three different vegetable crops found that nutritional content—protein, calcium, iron, vitamin C, magnesium, vitamin B2—declined anywhere from 6 to 38 percent from 1950 to 1999. And that was before glyphosate use fully skyrocketed. "You look at the rise of all sorts of different illnesses . . . and, well, what it comes down to is poor soil practices," Alex said. "We're just slowly killing ourselves. And the world."

"No-till" agriculture—an approach that holds promise for both reversing the damage we've done and restoring the world's soils—is a term now unfortunately rife with misconception. Alongside its original organic intention, it's being deployed in conventional agriculture, which is using genetically modified crops and copious amounts of glyphosate to reduce weeds without the plow under the guise of keeping carbon in the soil and minimizing erosion.

But for Alex and Yoko, no-till farming constitutes an approach that is regenerative in the truest sense.[3] "Our practice is really centered around soil health," Yoko said. They don't use a tractor and minimize disrupting the soil: Rakes and a broadfork are their most used hand tools; they take a pitchfork or the occasional shovel or trowel to turn compost; and they employ silage tarps to kill perennial weeds to prepare a new field. This approach keeps the ground's structure and ecosystem intact. They even abandoned growing Yoko's delectable burdock since the root system causes a massive disturbance when harvesting. "It's not no-till when you're digging out a four-foot root," Alex laughed.

Additionally, Assawaga Farm is certified organic, but Yoko and Alex avoid the organic pesticides, prepared compost, and plastic mulch widely used on other organic farms.[4] They also follow four out of the five principles of soil health pioneered by Jay Fuhrer, a soil health specialist with USDA's Natural Resources Conservation Service: keep soil covered; no tilling; encourage plant diversity; and maintain living roots in the soil. "The fifth [incorporate grazing animals] is kind of harder to do in our context," Yoko said.

Still, Yoko made it clear that their fastidious approach was borne of fascination rather than doctrine. "Some people are so annoyed

---

3 Although when I asked Yoko what she thought about the term *regenerative,* she had this to say: "I feel like the word *regenerative* is just another word that big agribusinesses are using to sell their products . . . so I feel like there's contention."

4 Plastic mulch is permitted by organic regulations, provided it's removed at the end of the growing season.

when they hear *no-till* because they think it's this super-dogmatic, black-and-white kind of thing, and it's not. Everybody's style of farming is different. And that's why farming is so interesting and fun." Case in point: Yoko's favorite subject growing up was biology. But on the farm, one day she could be a scientist, the next an artist: She made block prints and had started working with natural dyes from plants they were growing on the farm, like indigo. Farming, it turned out, was the ultimate outlet for her intellectual curiosity, creativity, and lifelong love of the outdoors.

Agriculture, after all, is an art as much as a science, and Assawaga was like a breathtaking exhibition and soil health laboratory: The rows of remaining fall vegetables were bordered by perennial hedgerows, including native pollinator plants and hazelnut, cherry, and crabapple trees. There was also a you-pick flower area, which with the hedgerows would support biodiversity both above- and belowground. Crop beds were carpeted with salt marsh hay sourced by "only one guy" from the marshes north of Boston to conserve water and minimize erosion. And the adjacent field was plush with a multispecies winterkill cover crop, which would die during the winter and leave a protective mulch to feed the soil. Yoko said she hand-mixed six different seed combinations, then listed each one's seasonal planting window like a mad scientist of cover cropping. "That's really important because every different species of plant feeds a different range of microbes," said Yoko, offering me a leaf of spinach to taste. "It's really meaty and sweet," she said.

I admired the glossy furrows, then folded it into my mouth. I hadn't tasted spinach that real since the beloved bacon mustard vinaigrette salads of my childhood (yes, I understand I was an odd kid by American standards); it seems these days you can only find the flimsy baby stuff, and in polluting plastic clamshells to boot.

"It's really crisp!" I exclaimed as half a leaf lodged in my throat, prompting a coughing fit. (I wasn't prepared for such heft.)

"Yeah, people go crazy for it," she said, then went with me to grab my water bottle.

Unbelievably, despite all of Alex and Yoko's environmental efforts, the real paradigm shift didn't occur for them until last year, Alex divulged. That's when they started making their own compost—a special variety sprinkled on the farm's soil to act as an inoculant, much like a probiotic capsule would for your gut. The process didn't sound all that exhilarating at first until Yoko explained another farm-sanctioned hand tool they were crafting it in conjunction with: a microscope. She and Alex learned how to use one to fine-tune soil health via a recent course with Soil Food Web School, founded by microbiologist and soil research pioneer Elaine Ingham.

"It's really, really cool to actually be able to take a sample of whatever bed and see under the microscope what's happening," Yoko enthused, describing their major garlic crop failure two years prior as an example. "We were a little ambitious. We were doing some crazy experiment with cover crops." But through the microscope, she and Alex were able to spy organisms that shouldn't have been present in healthy soil. As a result, they were immediately able to determine what went wrong: The soil had become anaerobic, or devoid of oxygen. (Generally speaking, belowground organisms, like we aboveground organisms, favor oxygen to survive.)

"When we talk about soil health, it's so theoretical, you're just kind of imagining things," she said. "But then, being able to actually see it under a microscope, it really brings it to life. You start to get more intimate with the soil and [its] inhabitants."

Alex was equally awed. "It really changed the way that I looked at the soil, how we grow food, how healthy that food is, how nutrient dense it is—and I think that's a big component that a majority of farmers, a majority of growers, just don't comprehend . . .We're just part of the process; there's other things at work. You know what I mean?" he laughed, an emotional release while contemplating the vastness. "It makes farming very deep for me."

And clearly, it wasn't only the soil benefiting; their business was too. "When we do the Brookline market, the market starts at 1:30. And at 12:45, we have a huge line. People get there early. We're the only farm they do that for," Alex said with so much pride that it didn't seem a bit boastful. "But there's a reason, you know? Everything tastes better, looks better . . . it's indicative of everything else that we do, our concentration on soil health." He walked over to a nearby bed and pulled out a vermilion carrot, then handed it to me. It was tender, with the terroir of sweet earth.

"One of our specialties that we'll start harvesting," he said. "It's a Japanese variety called Kyoto red."

"People start asking about our red carrots two weeks into the season," grinned Yoko.

FIVE MINUTES AFTER PULLING away from the tucked away paradise of Assawaga Farm, I passed a Walmart. Alex had sighed that most of the local community still shopped there, despite all of Assawaga Farm's success in Boston. It was a quandary I would witness again and again: Land prices too steep in urban areas to support upstart farmers, and communities in rural areas too ensnared by corporate colonization to support the visionary ones who came there. Hours later, before my flight home, I would grab dinner at a small-town Italian joint I once cherished. A woman berated the man behind the counter over the price of a pizza while a family sat eating gloppy pasta out of foil take-out containers, their children rolling around on the floor, gaming on iPads.

Not everyone in this world will find their way back to a rooted existence. But Yoko and Alex gave me hope for the ones who try. "I think the one way I see we can make an impact is just showing people how what we're doing works—and that it's profitable," Yoko said before walking me out of the barn beneath a chandelier of drying *takanotsume* chili peppers. "If that allows one more young farmer to [think], *Oh, I can do this*, and start their own farm—that's kind of awesome."

SIX

Black Snake Farm

"In the face of overwhelming odds, I'm
left with only one option: I'm gonna
have to science the shit out of this."

—Mark Watney, *The Martian*

ORIGINALLY, AARON BAUMGARDNER ENVISAGED HE MIGHT GO into medicine. So once in college at The University of Akron in Ohio, he joined the school's chapter of Phi Delta Epsilon, an international premedical and medical fraternity. For many of Aaron's fraternity mates, the health care calling was clear. "'I want to go to Africa and help [fight] some communicable disease,' or 'I want to cure cancer!'" he recollected their avowals in a goody-two-shoes voice. Aaron had alternate plans.

"I was like, I want to go and be a pathologist and I want to tell people in the courtroom why [the victim] died. . . . I wanted to say this person was bludgeoned to death by a baseball bat or whatever," he declared like the unseemly Kristen Wiig character in the Lawrence Welk sketch on *Saturday Night Live*. "So then I thought, maybe this isn't for me!"

Aaron, twenty-seven, is now the director of natural resources for the Catawba Nation, the federally recognized Native American tribe whose ancestral homeland extended from southern Virginia through the present-day Carolinas for at least six millennia.[1] He came to this work—his own calling, it would seem from our conversation—from his discipline as an ecologist, the alternate area of study he pursued.

---

1 Archaeologists have traced the arrival of humans in North Carolina—the ancestors of the Catawba—to the end of the Pleistocene, some 12,000 years ago. But considering emerging evidence that humans inhabited North America tens of thousands of years earlier than previously thought, the Catawba lineage to that landscape could extend further.

As it turned out, it was an interest in biology that had at first steered Aaron toward medicine. He also homed in on a passion for plants, rather than patients: Seeking lab experience to fulfill his pre-med requirements, Aaron wound up at a plant pollinator lab. There, he assisted studies on heterospecific pollen—the "wrong" pollen deposited on flowers that may not only drive an individual plant's reproduction but biodiversity itself—research that has profound implications for a planet where bees, butterflies, and other pollinators are being driven to extinction. "And that was my introduction that led down the rabbit hole of going out in the field and loving to be outside and just loving the natural world," he said. Growing up in Ohio, Aaron had rarely even gone hiking. But in college, "It really became my life."

I had phoned Aaron one November morning in advance of visiting farms in South Carolina over the week of Thanksgiving. Given the whitewashed narrative of that holiday—a story of helping and harvest that conveniently omits the subsequent genocide of America's original peoples—it seemed a fitting time to talk with an Indigenous American who could personally illuminate the inequitable state of American agriculture.

Today, white people own 98 percent of US farmland, while 54 percent of farmworkers are mostly undocumented Mexican immigrants. Meanwhile, Native, Black, and Asian Americans grapple with the legacy of mass land dispossession. Yet Catawba Nation, just outside Rock Hill, South Carolina, is at the forefront of an Indigenous movement to recover not only land but food sovereignty: In 2020, in the midst of the pandemic, the tribe started Black Snake Farm, a twenty-two-acre mixed-use farm now producing food for Catawba citizens. Aaron, who manages the tribal trust lands and the Nation's wildlife preserve, was tasked with overseeing the farm, though he had no training in agriculture. "I don't describe myself as a farmer," he said. "My background is in biology."

But what I would discover is that Aaron's own journey has been as convoluted as the path to Indigenous food sovereignty. "This is getting into the real thick of it, but I grew up Mormon," he divulged when I asked him how he felt about the coming Thanksgiving holiday. "And so if you know anything about Mormonism, they have this perception of what Native peoples are. They believe that Native people are Jews who escaped Babylon and got to the Americas. . . . So growing up, my perception of being Native was tied to being Mormon."

Being among the Mormon-condemned, I was aware of this origin story. In fact, Aaron first launched into his ancestry because I asked about the origin of his last name (Baumgardner) before I learned he had grown up Mormon. "So the last name is not Catawba," he laughed at the obvious, but then enlightened me: The name is German. His grandmother, who was Catawba, married Aaron's German-descended grandfather after her family moved from their ancestral land to Ohio in the 1940s. In Ohio, there had been a Moravian settlement of Delaware Indians where the Catawba joined up to "play-act Native American" in a historical village and do crafts, he explained. "Even down here, people are like, *Are you Catawba?* And I'm like, *yeah,*" sighed Aaron, who has dark brown eyes, dark brown hair, and a scruffy beard to match. (An inability to grow facial hair is a stereotype about Native Americans.) "And then I have to give them the whole family history for them to know, because Baumgardner is not a name that's relevant here."

Further illustrating the Catawba people's displacement, Aaron said the Catawba family name he is most closely related to is Beck, actually a Cherokee name. The Catawba had once been second to the Cherokee as the most populous tribe in the Carolinas, and the two fought over hunting grounds in ancient times. The two tribes took opposite sides during the Revolutionary War (the Catawba aided the colonists; the Cherokee, the British) but later intermarried in the aftermath of smallpox, forced relocation, and near extermination. "A lot of people here hyphenate to keep the Catawba name," Aaron explained, adding that only five or six last names encompassed the majority of Catawba households. "It's a very tight-knit type of community."

Aaron has had "a cultural reckoning" since his youth, he said, as has his sister, who recently moved to South Carolina and now works for the Catawba Cultural Center. The two of them are still "sort of figuring out" the beliefs imbued through their Mormon upbringing while also learning about their true culture. This could be the case for much of the Catawba community. "Fun fact: Catawba Nation is about 50 percent Mormon," he said, followed by an uncomfortable laugh. Mormon missionaries came proselytizing in the 1880s; by the 1920s, nearly all the remaining Catawba had converted to Mormonism. "We have a complex history."

FOR THE CATAWBA, WHO call themselves *yeh is-WAH h'reh*, or "people of the river," it has been a long road back to land sovereignty—the right to govern and steward their traditional lands along the Catawba River. Unlike many tribes of the Northeast, who first experienced European contact when English colonists began arriving in the 1620s, the Catawba's first encounter was almost a century earlier in 1540, when Spanish conquistador Hernando de Soto and his gold-seeking explorers traipsed through Catawba territory. (Evidently de Soto wasn't satiated after plundering the Inca Empire with Francisco Pizarro eight years prior.)

Then came encroachment, both human and viral, via settlers to the Carolina Piedmont in the eighteenth century. In 1763, the Catawba secured a title from King George III for 144,000 acres of their once far-reaching landscape, but this was only after their population had been virtually obliterated by smallpox: Of as many as 25,000 Catawba people a half century earlier, less than a thousand remained.

Later, in the Treaty of Nations Ford during the Trail of Tears expulsion—in return for being "allowed" to stay and the promise of a $16,000 payment plus a swath of secluded land—the Catawba ceded those 144,000 acres to the state of South Carolina. The Catawba fought for tribal sovereignty on what minimal land they were finally granted. (It is unclear from the record if the promised resettlement money was ever paid.) Their federal status was removed in the 1950s and then only rebestowed in 1993, after the Nation abandoned their claim to those 144,000 appropriated acres. The Catawba made the apt but ultimately unsuccessful argument that the treaty had not been ratified by the US government and was thus illegal. What transpired can only be described as land theft: Today, the Catawba reservation encompasses 700 acres near Rock Hill, South Carolina, just outside Charlotte, North Carolina.

Still, those 700 acres sit along the wooded banks of the Catawba River, where Aaron's ancestors lived for thousands of years. He is also leading a farm where his forebearers farmed for millennia, since the Catawba, among many Native peoples, were America's original sustainable agriculturalists—growing corn, beans, and squash to supplement fish caught from the river and the deer and wild turkey they hunted. Not many Americans in this age can claim that kind of antediluvian connection to a landscape. Aaron moved to Catawba Nation in October 2021, after first looking for jobs out West and stumbling across the director of natural resources job posting on the Catawba Nation Facebook page. "It's been a nice homecoming," he said.

IN THE HOMETOWN OF his youth, Akron, Ohio, Aaron and his family frequented museums and historical tours. Before medicine, he thought about zoology, although he had no role models for a career in the sciences, save for an aunt who worked in a cancer biology lab. His parents are high school–only graduates: Aaron's mother works in a medical parts factory, and his father was in construction and maintenance. "They instilled the idea [to] do whatever I wanted to do and try to be as educated as possible. Luckily it worked out in my favor," he said, adding that it opened a lot of doors. "But I also think that education, for the most part, is a scam. That's what I'm kind of learning," he said half-jokingly. He could have been alluding to the cost of higher education, the omission of post-1890s Indigenous history from US textbooks, the Mormon indoctrination, or the area of work at Catawba Nation for which he received no formal schooling: farming.

His education, however questionable to him now, had been as a scientist. After finding his way to plant ecology at the University of Akron, Aaron headed to grad school at California State University, Bakersfield, where he studied the impact of drought on Southern California's chaparral plant communities and ventured to the ocean, desert, and sequoias in his spare time. Later, he transplanted to Kansas, where he conducted research on plant ecology and population biology at Kansas State and the University of Kansas and witnessed the detriment of mega grass monocultures like corn on the native prairie ecosystem.

But like singularly qualified astronaut-botanist Mark Watney in the sci-fi survival novel-turned-film *The Martian*, it turned out plant ecologist Aaron was the perfect pick to guide the Catawba effort to start a farm and grow food. Ironically, the land he was handed may as well have been Mars: It was red clay, notoriously dense and challenging, with near zero topsoil. "One of the silly decisions of a previous manager was to strip the entire topsoil off the area they were going to be cultivating," Aaron scoffed. He was given no logical explanation as to why it was removed. "I don't know why they didn't just amend the existing topsoil." And he, like Watney, would be improvising as he went along: "I don't have a background in farming. My farm manager doesn't have a background in farming. And none of my staff have backgrounds in farming," he said. (There are three full-time staff members, two of whom were supposed to be part-time before realizing the demands of the undertaking.)

Yet as Aaron detailed how they're growing food in such depleted soil, and restoring it by incorporating principles of ecological succession, it was obvious he was the right person for the job. They are planting cover crops and trees, bringing in carbon and compost to engineer a lasagna bed system, and minimizing tillage to recover microbial health. He was as insightful as any veteran sustainable farmer I had interviewed. I told him as much.

"You know, I see agriculture as just an organized ecosystem," he laughed. "And so that's why, when I started managing the farm, I [said], 'We're going to move toward organic regenerative agriculture; we're going to try to till the land as little as possible.'"

Interestingly, not everyone at Catawba Nation saw the value in those tenets of sustainable farming. I asked Aaron what he saw as the biggest obstacle, and he didn't mince words: "They see it as some hippie stuff," he said. The word *organic*, too, was problematic. "They think that it's hoity-toity. Or they don't necessarily understand what *regenerative* is . . . and so I think it's really about trying to figure out how to best educate people [with] this return to these principles that have this new name but would have been very familiar to our ancestors."

The pushback Aaron reported had one obvious explanation: Native Americans are Americans. Many have grown up in as un-regenerative an environment as anyone else in modern-day America. But I have to admit I was surprised about the contention, considering that in my fifteen years covering the environmental movement, Native Americans were perpetually described as the "original environmentalists." And yet this stereotype—the "ecological Indian"—had insidious origins, I discovered as I dug into the research. It was formed by European settlers, who regarded Native people as subhuman wilderness creatures, then used that delusion to justify the taking and civilization of their "untamed" land. Later, John Muir's misconception of uncovering a "great, fresh, unblighted, unredeemed wilderness" in Yosemite and elsewhere fueled the founding of the National Park Service, which sparked the modern conservation movement.

"In other words," writes activist Dina Gilio-Whitaker, a member of the Colville Confederated Tribes, "humans were viewed as separate from, and even a threat to, a pristine natural environment. Yet indigenous [*sic*] peoples hadn't just lived sustainably in virtually all of the landscapes on the continent for thousands of years; many Native nations are also known to have had complex land management practices." The majestic natural resources European settlers and Muir witnessed were *because* of human involvement, not in spite of it. "That these facts were and are systematically ignored was part of larger patterns of erasure, genocide, and dispossession."

Aaron echoed these revelations. "You know, our ancestors weren't just foraging, but they were doing agriculture within the Carolina Piedmont. They were lighting fires to maintain a savanna prairie ecosystem that would have been more abundant with things [than] they would have eaten [from] a forest. And so they were actively managing the land in a way that the land needed them, and we needed the land."

⁓

WAYS OF LIFE ARE ever evolving, of course. "We're not trying to go backwards," Aaron noted. Yet returning to a reciprocal relationship with the land is an urgent matter. Catawba Nation purchased the land for Black Snake Farm at the onset of the COVID-19 pandemic, in 2020, using relief funds from the CARES Act.[2] "I think there was really a citizen need and then also, from Catawba Nation's standpoint, the realization that the global food system was fragile," said Aaron, who took his position in 2021. "We all saw that when we went to the grocery store."

Beyond food security, having access to fresh, unprocessed food is also critical to the health of the Catawba people: Diabetes is more prevalent among American Indians and Alaska Natives than any other racial group in the United States, nearly double the rate of non-Hispanic whites. The CDC predicts that one in three Americans will have diabetes by the year 2050. That statistic is for the general population; for Native populations, the disproportionate increase would be unfathomable.

Encouragingly, the call to reclaim Indigenous food systems is amplifying. Along with Black Snake Farm, the movement now includes organizations such as the Native American Food Sovereignty Alliance, home of the Indigenous Seed Keepers Network; North American Traditional Indigenous Food Systems, founded by Sean Sherman, the James Beard Award–winning food activist known as the Sioux Chef; the Natwani Coalition, which supports Hopi and Tewa growers in Arizona; and the intertribal Dream of Wild Health outside Minneapolis–St. Paul, which runs a thirty-acre organic farm, an Indigenous food share CSA, and the Garden Warriors training program for Native youth.

Black Snake Farm's first season of production was 2021. By the end of their second season, Aaron and his team had created three

---

2 It is a preposterous throughline to consider: Catawba Nation had to apply to the US government for money to pay a private landowner for land stolen from the Catawba by the state of South Carolina, abetted by . . . the US government.

and a half acres of enclosed gardening space and were also tending approximately twenty egg-laying hens, two honeybee hives, and three steers. (Aaron's designated farm manager didn't have vegetable farming experience but grew up raising livestock.) All in all, they had grown three-quarters of an acre of vegetables; planted a half acre of trees donated by The Fruit Tree Planting Foundation; and devoted another quarter of an acre each to traditional "three sisters" crops—beans and squash in one field and, in the other, melons with their recently rematriated Catawba flour corn.

Aaron estimated that year they provided vegetables and eggs to seventy Catawba households—primarily seniors. Additionally, his team distributed beef to over 200 families with just those three steers. "That was giving them a package of hamburger or a few items here and there," Aaron qualified. "But we're growing. And we're trying to better figure out how many people we're truly serving—and what we're even capable of serving," he said. Most recently, Catawba Nation launched a weekly farmers' market.

Aaron envisions that one day the farm could provide traditional foods to thousands of tribal households. Catawba Nation currently has 3,000 enrolled citizens, but Aaron estimated there are some 2,000 within the farm's immediate reach—either living on the reservation or in the vicinity of Rock Hill and other nearby cities. "For us, [the farm] is really about sovereignty as a nation," he said. "And food sovereignty is very much connected to our sovereignty as a Catawba people, right? Food is one of those basic needs . . . to be able to survive. And if we, as a nation, are unable to provide food to our people, then it makes me question whether we really are a sovereign nation."

For the immediate future, though, Aaron was realistic. The goal, he said, is not isolationism or never being able to shop at Walmart. But as we all intimately know post-pandemic, the world is uncertain. "If we had to become insular and provide for our community, provide for our families, I want to make sure that we have enough seed and enough food to be able to provide those needs. And so that's what I'm working toward, even though it's a very lofty and probably many decades-long goal."

Aaron's realm isn't limited to agriculture. In the long term, he said, he hopes to restore the Piedmont savanna and its wildlife, as well as steward resources for the Catawba to continue their traditions of fishing and hunting. "I really have found a rewarding career working for my tribe," he said. "Not that the science that I was doing at . . . other institutions wasn't meaningful. It was just a lot harder to say, *This bit of knowledge will eventually help someone.* Whereas right now, doing the work at Catawba Nation, I'm seeing the difference in the work my staff and I are doing—being able to supply people with meat, or vegetables, or eggs, or corn, and seeing that it's going to be on their table."

As for Aaron's own table that Thanksgiving we spoke, he said the vegetables from Black Snake Farm would be reserved for Catawba's more needy seniors, but he saved a bag of grits and cornmeal from his corn "babies" he had been caring for all season. He and his team had nixtamalized the crop and ground it into different sizes—cornmeal, masa, and grits. (Nixtamalization is the traditional process of soaking maize in an alkaline solution to render it more digestible.) Aaron planned to transform it into cornbread stuffing to celebrate the holiday that year. He surmised most people within Catawba Nation would also celebrate the holiday, but then added a caveat. "My sister and I are going to take a different approach this year and really try to be a bit more transparent about what we're being thankful for. Because I think it's still important to be grateful during this harvest season and have that reflection time, but really be more truthful about the storied past."

Still, there would be a turkey, and dressing, and homemade noodles on top of mashed potatoes from his Amish Country, Ohio, upbringing, alongside his farm-grown stuffing. And the life of the corn they grew wouldn't end there: In January, he planned to teach Catawba citizens how to make tortillas—a simple way to empower folks to eschew store-bought convenience. "Not that [tortillas] would have been something [the Catawba] would have probably used regularly," Aaron laughed before revealing the source of this adopted tradition. "My husband is Mexican. . . . We have his grandpa's tortilla press. I'm the one that uses it mostly. I do love making some fresh tortillas whenever possible."

"Culture and tradition aren't stagnant," Aaron had wisely reminded me earlier. Real, delicious food, however, is ever enduring.

SEVEN

FarmaSis

"Our disconnection with food, our
disconnection from one another, is by design."

—Germaine Jenkins

The breadth of one acre at Mahonia Gardens, a market garden in Sisters, Oregon. *Photo by Benji Nagel.*

Farming intern Elliott Blackwell replenishes The Stand, Mahonia Gardens' self-serve farm stand in downtown Sisters. *Photo by Tai Power Seeff.*

Mahonia Gardens founders Benji Nagel and Carys Wilkins in front of their home after the July garlic harvest. *Photo by Tai Power Seeff.*

Lunch break at the farm: Carys's mom, Maribeth Quinn, plays with Carys and grandson Junius in her kit-built wooden yurt home. *Photo by Tai Power Seeff.*

The outdoor kitchen at Wild Abundance, a school for permaculture and homesteading outside Asheville, North Carolina. *Photo by the author.*

Wild Abundance founder Natalie Bogwalker and permaculture apprentice Dan Hancock show off a harvest of Chassa Howitska, or Seminole pumpkin. Natalie's hand-built cabin is visible in the background. *Photo by the author.*

Carpentry instructor Alex Kilgore assists a student with a drilling lesson during Wild Abundance's four-day Women's Basic Carpentry class. *Photo by the author.*

Farmer Matt Oricchio collects seeds off an amaranth plant at Ambler Farm in Wilton, Connecticut. *Photo by the author.*

Ambler Farm's director of agriculture, Jonathan Kirschner, admires robust artichoke plants in the farm's perennial field. The crop that looks like corn is actually sorghum-sudangrass. *Photo by the author.*

Parents and kids begin arriving for Ambler Farm's Fright Night, a community event of pre-Halloween fun for third- through fifth-graders. *Photo by the author.*

The token herd (from left: Coco, Ven, Maggie, and Val) at Round Table Farm in Hardwick, Massachusetts. Farm owners Marlo Stein and Archer Meier source raw milk for their cheesemaking from other local dairies. *Photo by the author.*

Marlo waits for me to put on a hairnet before escorting me into the "make" room for the farm's aged raw milk cheeses. *Photo by the author.*

Archer nuzzles an extra-affectionate member of the farm's new herd of Nubian, Saanen, and Alpine goats. *Photo by the author.*

Yoko Takemura and Alex Carpenter at Assawaga Farm, their no-till organic vegetable farm surrounded by woods and wetlands in Putnam, Connecticut. *Photo by the author.*

Yoko and Alex's hand-built barn/residence overlooks vegetable beds surrounded by perennial hedgerows; in warmer months, this border will spring to life with more than twenty-six species of native pollinator plants. *Photo by the author.*

"People purchase with their eyes," says Yoko. Vegetables are artfully displayed beneath garlic and dried flowers at Assawaga's Saturday (formerly Sunday) farm stand. *Photo by Yoko Takemura.*

Volunteers husk Catawba flour corn during Catawba Nation's corn harvest event at North Carolina's Davidson College. The school is collaborating with Catawba Nation's Black Snake Farm to grow the tribe's recently rematriated corn. *Photo by Lea Newman.*

Catawba flour corn; the seeds had been lost to the Catawba people for more than half a century. *Photo by Lea Newman.*

Lindsey Barrow and Olivia Myers of Charleston's Community Supported Grocery pose with Nell, their school bus turned sliding-scale mobile farmers' market. *Photo by the author.*

Third-generation Gullah Geechee farmer Joseph Fields gathers a handful of green onions at his eponymous farm on Johns Island in Charleston, South Carolina. *Photo by the author.*

Carolina Gold rice ready for harvest. The reintroduced heirloom crop was first grown in South Carolina's Lowcountry during the Revolutionary Era by enslaved West Africans who had vast knowledge of rice farming.
Photo by the author.

The women of FarmaSis. Founder Bonita Clemons is in the red boots.
Photo by Anthony Montgomery.

Grazier and Grass Nomads co-owner Ariel Greenwood: "I grew up rurally and we had horses. It's the substrate I'm happy in." *Photo by Sam Ryerson.*

Ariel's husband, rancher and Grass Nomads co-owner Sam Ryerson, moves cow-calf pairs to new pasture at 7,000 feet in northeastern New Mexico. *Photo by Ariel Greenwood.*

There aren't a lot of couple selfies when you're both on horseback: Ariel snapped this photo of Sam as well as the one above. *Photo by Ariel Greenwood.*

Free-range kids follow a nature trail past "animal village" in
the agrihood of Serenbe, outside Atlanta. *Photo by the author.*

A weekly CSA pickup at Serenbe Farms. *Photo by the author.*

The land outside Arlington, Washington, that will become permaculture farming community Rooted Northwest. *Photo by Rooted Northwest.*

It takes an agrivillage: Rooted Northwest members build their first greenhouse. *Photo by Morgan Keuler.*

Farmer and food educator Alex Rosenberg-Rigutto plants garlic from homegrown seed at her Northwoods Farmstead and Skill Center. The harvest earlier that summer (the farmstead's first) yielded more than 1,800 heads. *Photo by Alexandra Rosenberg-Rigutto.*

An Ashkenazi superfood, cabbages are harvested and trimmed at Tamarack Camps' Farber Farm. *Photo by Alexandra Rosenberg-Rigutto.*

Farber Farm's winter squash and pumpkin haul. Varieties, clockwise from front left: Black Futsu, Delicata JS, Waltham Butternut, and Winter Luxury. *Photo by Alexandra Rosenberg-Rigutto.*

THE ELEPHANT IN THE ROOM IN CHARLESTON, SOUTH CAROLINA, is that the city is sinking. As our plane descended through storm clouds, I peered out the window at marshland riddled with rivulets like veins and heard *The Rime of the Ancient Mariner*: "Water, water, everywhere . . ." I imagined one rolling, monster wave, and it could all disappear into the ocean. I reached across the aisle and tapped my husband on the shoulder, then nodded toward the window. "Oh yeah, nothing to worry about here," he joked, then glanced to a lineup of Lockheed C-130s, no doubt on standby after Hurricane Ian that fall. Later, driving to my mom's house at dusk, I gazed past concrete pilings half-submerged in pluff mud, harbor lights blinking in the distance.

Before my mom and stepfather moved here six years ago, I—helpful daughter that I am—sent them the city's *Sea Level Rise Strategy,* now more aptly named *Flooding and Sea Level Rise Strategy,* which contrived how to cope with a projected fourteen to eighteen inches of sea level rise in the next twenty-five years. This, on top of the more than a foot of rise that transpired in the past century, plus the compounding problem of subsidence (the land itself is really sinking) that scientists are racing to understand.

"The house we're looking at is in a high area," my mom said, and it is, except: The main road that leads to this lofty neighborhood traverses wetlands and sometimes becomes impassible due to king tide flooding—not to mention tropical storms and hurricanes. Back in 1995, Charleston averaged eighteen of these "nuisance" flood days, which turn downtown into a deluge. Twenty years from now, the city will see 180 of them yearly, or one every other day. Every new construction house I saw was built on an elevated foundation at least a story high. This uplift may soon be required by law, and there were *a lot* of those gawky houses: Charleston is on the precipice of climate catastrophe, but it is also one of the fastest-growing cities in America, with half a million new residents expected in the next thirty-five years. Meanwhile, one of the thoroughfares I traversed all week is called—I kid you not—Folly Road, which just about sums up the situation.

Doom and gloom aside, Charleston appeared to be shoring up food security for the future, with new young farmers raising everything from pastured poultry (Jeff Siewicki at Vital Mission Farm), grass-fed beef (Alex Russell at Chucktown Acres), and oysters (Josh Eboch at Barrier Island Oyster Co.) to luffa sponges (Brian Wheat at Common Joy; his wife Alison Pierce farms oysters at Barrier Island), sea beans (Sam Norton at Heron Farms), and a tropical food forest (Anthony Natoli and Ellen Waldrop at Fire Ant Farm). This, along with spending Thanksgiving with my family, was why I had added carbon to the atmosphere to travel there.

Not surprisingly, the city has growing renown as a farm-to-table foodie town. Wasn't that why people were flocking there (along with the balmy weather and beachy lifestyle)? The region seemed to be reclaiming a local food culture.

"If you're white!" said Nikki Seibert Kelley, the director of Growing Local SC and founder of consultancy Wit Meets Grit, when I relayed this observation to her before my trip. Nikki got her start as a mentee of Joseph Fields, a third-generation Gullah Geechee farmer stewarding fifty acres on Charleston's Johns Island. The Gullah Geechee people are the descendants of enslaved West Africans who maintain an intact culture and Creole language in the Southeastern United States. "That's how I really got immersed in farming," said Nikki, who is white, referring to her time working with Joseph. "But I also inherently got a lesson in how Black farmers are perceived and treated."

First, there was the legacy of slavery. I shouldn't even have to point this out, but this calamity was long genteelly assuaged and even romanticized in Charleston (think Old South tours and plantation weddings), which contributed to the area's "toxic race culture." Then there was the violence and institutional racism after abolition, including USDA defunding beginning in the 1950s, which drove Southern Black farmers from 12 million acres of their land over the past century. Today, inequity extends from the area's farmers to eaters: Agriculture may be South Carolina's biggest industry, to the tune of $50 billion a year, but the state exports 90 percent of its food. The local food that remains and lends the town its farm-to-table reputation mostly benefits a privileged group of restaurant-goers.

"Tourists are eating up everything here," Lindsey Barrow told me. He and his partner, Olivia Myers, are the founders of Community Supported Grocery, a sliding-scale local food delivery service that vends online and out of a green school bus turned mobile farmers' market. When I met the dynamic duo the morning after I arrived, he described the disconnect. "Every year, we win best [food] city in the country, best city in the world, there's James Beard Awards coming out, blah blah blah. But there's data saying just a little ways up the road, children are living further beneath the poverty line than anywhere else in the state of South Carolina."

It was time to unpack not just Charleston's, but agriculture's other elephant in the room: racism. Plus, the ultimate force muddying the regenerative farming conversation: capitalism. Without confronting both, the small farm movement doesn't stand a chance.

MY SECOND DAY IN Charleston, my nostrils burned from the fumes of an industrial paper mill I passed on the way to one of those "just up the road" places: the Chicora-Cherokee neighborhood of North Charleston, where the average household income is less than $19,000. There, Germaine Jenkins, who attended culinary university Johnson & Wales as a single mom in her twenties, founded Fresh Future Farm in 2014, a nonprofit community farm and sliding-scale grocery store on a once-vacant city lot. The 0.81-acre farm will soon be expanding to a twenty-acre rural site.

Germaine is now fifty-two, with two grown children who are farming at Fresh Future and "farm baby" grandchildren who love gobbling up the site's blueberries, bananas, blackberries, peaches, pears, figs, and loquats. She may also be more famous by the time you read this: When we met, she had recently wrapped five years of filming for *Rooted,* a documentary following her journey to bring fresh fruit, vegetables, and food (i.e., food justice) to a community where the last grocery store folded in 2005.

Germaine, wearing a Fresh Future Farm merch sweatshirt and bright yellow sneakers, didn't say a word as she unlocked the chain around Fresh Future's entrance and led me past pecan trees into the bodega. And she was soft-spoken when we first started talking—me in a folding chair, her behind the cash register. But when I asked Germaine what she most wanted me to convey, she didn't demur: "It's not a mistake that this place is disconnected from the land," she said. "Our disconnection with food, our disconnection from one another, is by design."

Communities like Chicora-Cherokee[1] are often described as

---

1 To add to the layers of irony: The neighborhood is named for the Native peoples originally disconnected from this land.

food deserts, regions where people lack access to healthy, unprocessed food. North Charleston, which encompasses seventy-six square miles, has eleven of these bereft areas.

But the more apt descriptor for these regions is *food apartheid*, a term coined by urban farming activist Karen Washington. Along those lines, Germaine explained the systemic orchestration in Charleston and cities across the country to isolate Black communities as repercussion for the civil rights movement, giving the West Ashley neighborhood south of there as an example. "That community used to be the Black Wall Street of Charleston, [with] Black hospitals, Black libraries, Black grocery stores, myriad Black businesses, and Black community of varying socioeconomic levels, all living together and taking care of each other," she said. That is, up until educator and activist Millicent Brown, who lived in that neighborhood, worked to desegregate Charleston's public schools in 1963. The Crosstown Expressway was opened in 1968. "[It] went straight through her living room," Germaine said.

The consequence of nutritional isolation, logically, was Black people's health. Fast-food chains infiltrated, peddling and pointedly marketing cheap, processed food. Food banks followed, handing out more of the same. Obesity, diabetes, high blood pressure, and other food-related illnesses became epidemic among Black Americans. "You know, there was a time when there were enough Black farmers that Black people did not have a need for food [aid], and as that went away, we moved to a system where the nonprofit industrial complex is supposed to feed us," Germaine sighed. "And we're still sick, no matter how much money or pounds are distributed every year."

Germaine said more municipal support for both urban and rural farming would go a long way. "Why not fund the work of making people healthy? It's way cheaper than a lot of these disposable solutions. We've had customers who have had dramatic shifts in health eating the food we grow and the stuff we sell that's not normally available in Black and brown communities." One of those community members is Germaine's own child, Anik Hall, now twenty-three and the farm's creative director, who overcame severe food allergies and recovered from blindness in one eye after being nourished on Fresh Future's food.

But the ultimate answer, for Germaine, was and is clear: Black farmers, once again growing food for their own neighborhoods. "That's the only way it's sustainable," she said. Earlier, she had reminded me that sustainability, at its core, is staying power "when you don't have a lot of resources."

I had to head to my next interview, and Germaine had to drop off some homemade Thanksgiving meals (turkey, stuffing, green beans, cranberry sauce, and a chocolate dessert) for some bachelors, but she left me laughing with this missive for those of you interested in making lasting change: "Put the book down, grow [your] own food, and then pay for Black farmers to do their thing."

FIFTEEN MINUTES LATER, I pulled up to the scene of the most clichéd smash cut I could have picked to drive home the gentrification worsening Charleston's food apartheid: Whole Foods. The 40,000-square-foot store, the second in the city, opened in the aforementioned razed West Ashley neighborhood in 2018. As I drifted through the sliding glass doors and toward the café, I scanned last-minute Thanksgiving shoppers piling their carts with clam-shelled pies and charcuterie platters, then passed the poke stand.

I hadn't chosen this loaded meeting spot; Bonita Clemons had. I would challenge any writer to fully encapsulate her vivacious spirit or the breadth of what she does, but I'll describe her first as the founder of FarmaSis, an all-Black, all-women sustainable farming incubator based in Columbia, South Carolina. "Even now, I still don't know what I want to be!" laughed Bonita, age fifty-seven. "When people say, 'Bonita, what do you do?' I'm like, *How much time do you have?*" Later, she nailed it: "What do I do? I do my best to show people how to smile on the inside."

If Germaine is envisioning a country renewed with Black farmers, FarmaSis is a model by which to achieve it. The concept is simple yet radical: Bonita mentors a group of ten women as they learn how to grow food; later, the women start their own groups and each group passes on those skills to ten more women. The program is in its nascence, but the five-year vision is to give rise to 10,000 Black women farming to feed themselves and their communities. "I tell people I'm creating an army of women," Bonita said.

I had been moved by a documentary short Bonita sent me about FarmaSis. And I read how she mobilized a farm-to-food bank effort through Axiom Farms Cooperative, her food hub, that fed more than 10,000 South Carolina families during the pandemic.[2] Still, waiting in the café for Bonita, I was a bit uneasy. There had been a scheduling snafu with Germaine, another farmer had canceled the day before, and then Bonita had asked to make a last-minute change before deciding to drive three hours round trip from Columbia the day before Thanksgiving. There was a Black-owned farm Bonita wanted to show me, yet by her earlier cryptic correspondence I came to understand this farm's name and location would only be revealed upon our meeting. As I considered the possibility that Bonita might be yet another visionary farming activist to flake on

---

2 Whereas farmers' markets allow farmers to sell directly to customers, food hubs allow small- and mid-sized farmers to get their food and products to larger wholesale customers (e.g., schools, hospitals, and retail chains). If you want to join the local food movement and don't want to be a farmer, food hubs are an exciting area to get involved in, given the increasing precariousness of global supply chains.

me, I readied to head back to my mom's, crack open a beer, and start my mashed potatoes.

What I didn't know is that once the unstoppable Bonita Clemons gets an idea in her head, the outcome may as well have materialized.

Bonita appeared right on time, wearing a pom-pom beanie, thin gold hoops, and a multicolored fringed cardigan over a FarmaSis logo T-shirt. Bonita shook off my worries with one jubilant hug, then divulged the reason for the needed schedule change: Yesterday, she had been tied up donating collard greens, sweet potatoes, and cabbage to individual homes and The Uhuru Academy, an African-centered private school run by two FarmaSis grads. Then, to unknowingly tip me off more to her can-do, generous nature (and make me feel worse about having sat there, sighing for beer), she gave me the rundown of her "big, melted family" back in Columbia, which includes three nieces she co-parents and another niece and three-year-old grandnephew who live with her.

Her family undoubtedly incubated her gift as a connector of multitudes. "My grandmother had eighteen children. I'm the oldest of—we all lost count, but we know it's over fifty grandkids and eighty great-grandkids," beamed Bonita, who with her glowing skin and lithe build didn't look a day over forty-five. "She's ninety-five and still working."

"Wow! What does she do?" I asked.

"She's a crossing guard. And it's a four-lane highway," she said.

"At *ninety-five?*"

"See, it petrifies me. The police officers don't even want to do the job! But . . . that's what she loves; who am I?" Bonita gave an infectious laugh.

Bonita, like her grandmother, has had to be fearless to find the joy. Bonita's mother died of a stroke at the age of forty-two, when Bonita was in her early twenties. Not long after, Bonita was in a car accident that left her transitorily disabled and warranted several surgeries. She recovered her health through meditation, yoga, and a vegetarian diet, went on to obtain a master's in public health from the University of South Carolina, and hoped to join the Peace Corps. "Then my sister was murdered," she said. I looked at Bonita, speechless.

"So I inherited a little girl. She was in the third grade when her mama was killed. So I couldn't go—didn't go—into the Peace Corps.... [But] raising her was the best decision I did ever, ever in life. Like the best." (This is the niece, now thirty-five, who with her little son lives with Bonita.)

To support herself and her family, Bonita has worked dozens of jobs over the years. She studied economics at Benedict College, then worked as an internal auditor at a bank and later as an accountant at a prominent Black law firm in South Carolina before graduate school—both remarkable choices, considering she first questioned the social construct of money at the age of five. ("I remember asking my dad: What is this thing called money? Why don't they just print enough for everybody?") Later, Bonita would fulfill her long-held dream to work overseas, traveling to Afghanistan as a Morale, Welfare, and Recreation specialist for the military.

But alongside it all, she pursued channeling her personal revelation around food, health, and wellness into serving her community. In 2001, she founded the nonprofit Dianne's Call, named for her mother, to provide programs to combat food insecurity. As Bonita explained it, Dianne's untimely death had been the result of chronic high blood pressure and "a very tough life" working in a factory and raising four children after splitting from Bonita's father.

But undoubtedly, Dianne's death was also the outcome of the grave health injustices Black people face in this country, which includes the theft of land from Black farmers and the intentional food apartheid Germaine described. But there is also extant medical racism that originated from slavery-era experiments positing Black people could withstand more pain, which still impedes access to adequate and unbiased care. Then add in, though there's no way to scientifically measure it—not that we need to, to acknowledge it—the epigenetic imprint of centuries of trauma.

Black women most bear the brunt of these health inequities: Four out of five Black women are obese or overweight, more than any racial group in the United States, which increases their likelihood of heart disease, stroke, diabetes, and thirteen (proven) types of cancer. Black women also have the highest rate of maternal mortality and are three times more likely to die from a pregnancy-related cause than non-Hispanic white women. During the COVID-19 pandemic, these afflictions led to calamity: Black women perished at a greater rate than any other group, male or female, except for Black men.

This was the context (minus COVID) in which Bonita first came to Sumpter Cooperative Farms back in 2005. SCF is a Black organic farming cooperative founded by Azeez and Fathiyyah Mustafa in Sumter, South Carolina. Their cooperative of now sixty farmers was, and still is, an anomaly: In 2003, SCF became the first certified organic produce grower in the state—at a time when less than 1 percent of the nation's farmers were Black. Today, 1.2 percent of all American farmers are Black,[3] and being both a Black and organic farmer is truly extraordinary: Just 0.5 percent of certified organic growers are Black in a country where only 0.5 percent of farmland is even certified organic.[4] To get even wonkier and put the audacious

---

3 A hundred years ago, before dispossession, 14 percent of American farmers were Black.

4 It's important to note that many more farms use organic growing practices than are certified organic. Organic certification is too expensive for many small farms, although this will hopefully change with the USDA's reinstatement of the Organic Certification Cost Share Program, which reimburses producers up to 75 percent of the certification cost. Still, I have to wonder: If the US government truly wanted to support small farmers, preserve the environment, and protect people from pesticide exposure, shouldn't it simplify the process and make organic certification flat-out free? Meanwhile, government subsidies to producers of commodities such as corn and soy average $16 billion a year.

mission of FarmaSis in perspective, Black women only operate 25 percent of this 0.5 percent.

Bonita traveled to SCF to buy some pesticide-free vegetables. She ended up staying three hours. "[Azeez] just talked, this educated man. And I'm like, *this is a long time!*" But at the end of their one-sided conversation, Azeez had a message for Bonita, which in her entertaining retelling sounded like the mystical start of renowned mythologist Joseph Campbell's hero's journey, simultaneously the Call to Adventure and the Refusal of the Call:

Azeez: *You're the one.*
Bonita: *Mr. Mustafa, I don't want to be the one.*
Azeez: *It's not your decision.*
Bonita: *OK, I'm the one to do what?*
Azeez: *You're the one to get the information out about the food.*
Bonita: *But I don't want to be the one. I just want to come here, get my food, and go back home.*
Azeez: *Well, too bad, lady. You're gonna be. You're the one and you talk back too much.*
Bonita: *I'm sorry for talking back. But I'm not the one.*
Azeez: *Yes, you are.*
Bonita: *OK, then tell me why.*
Azeez: *Because you have the piece of paper.*
Bonita: *What piece of paper?*
Azeez: *You have a master's degree. And people will listen to you.*

So Bonita kept going back—resolutely for the organic vegetables and more reluctantly for Azeez's three-hour food and farming lectures. One day, accompanying him to a church workshop where she delighted in doing a collard greens salad demo, she realized, *You know what? I'm the one! He done tricked me!* That Monday, Bonita called Azeez to see when she could stop by to share the video she made about SCF that they planned to play at the farm's festival that weekend. "His daughter-in-law got on the phone. She said he passed away. . . . I was so devastated. I just couldn't believe it."

The family held the festival that weekend in his honor, and Bonita showed the film. "[Mr. Mustafa] planted the seed [for FarmaSis] in my mind and in my heart," she said. "That's when it really started."

NOT LONG AFTER, SOUTH CAROLINA'S Clemson University put out a request for proposals for a farm incubator program at its Sandhill Research and Education Center, a fifty-acre green space helmed by the College of Agriculture, Forestry and Life Sciences. The school had prepped eight acres with cover crops and would offer plots to start-up farmers for three years, at an annual fee of $350, to kick-start their farms. By that point, Bonita was helping other Black farmers in the area plant and harvest, and enjoying the work—though much of her growing knowledge, she said, came from her granddaddy. "What I saw growing up: You plant a seed and it grows. People make it complicated," she chuckled.

Bonita wanted to marry this ancestral Black farming knowledge with the technology and financial resources of Clemson, with its more than $1 billion endowment. She also didn't want to do an incubator program on her own. So she put up a flyer calling for a very particular group of growers: "Black women who were dancers, artists, musicians interested in gardening and planting." Azeez had told her plants love the vibration of the music. "I wanted someone to be beating drums [in the field], and I wanted to have a good time."

An artistically inclined group of eleven women materialized—mostly in their sixties, including an herbalist, a hairstylist, a chef, and an elementary school teacher. Perhaps as a wink from the universe to the power of movements to exponentially multiply, there were twin sisters in the group and two additional women who each had a twin.

Then Clemson delayed the program. So for more than two years, Bonita and the women planted in each other's backyards, sharing the harvests and eventually saving and swapping seeds. They grew zucchini, watermelon, cabbage, collards, corn, kale, okra, cherry tomatoes, and sunflowers, among the bounty; they hosted plant-based potlucks and tested recipes like radish hummus; they shared stories and sisterhood.

The original plan had been to work with each group for five or six weeks; Bonita's first group stayed together for five years. "After the third year, they didn't want me to get more new groups. They loved each other. We're a family; we bonded. So after the fifth year, I was like, *Look, I gotta kick y'all out of the house.*"

Now with her second group and in the Clemson incubator space, the women are younger dynamos: One is a geneticist with a PhD. Another is a political organizer. The youngest in the group went to school for horticulture and already has a farm herself. Bonita has decided that the trainees in the group to follow will be required to bring a young girl along. "We gotta transfer the knowledge," she said. "It's getting lost."

Bonita has also streamlined the path to those 10,000 Black women farmers. Each group of ten trainees will work for one growing season: nine months, from January to September. (South Carolina's growing season is actually year-round, but I'm guessing even the tireless Bonita occasionally needs rest.)

"The metaphor is like when a woman has a baby. For the first three months, you get your foundation: We talk about soil. We talk about the body. Because FarmaSis has three goals: The first is collective work. It's really easier when we work together. Number two is health and wellness. And number three is economic development," she said, burying the lede. "Growing food is a hobby. Farming is a business. So this will let you know if you want to take it to the next level."

⚡

THERE IS HEALING IN working the land. I have found it myself, and it is what launched me on this journey. I've witnessed it, too, in countless others: like Germaine, whose child can see out of a once-blind eye and who has brought sustenance to a broken community; like Bonita, who channeled personal anguish into empowering a legion.

And when I followed Bonita out to Johns Island to visit that farmer she wanted me to meet, it was hard not to feel the magic as I drove for miles beneath intertwined oak and Spanish moss, and the land finally unfurled into the fifty acres of Joseph Fields Farm. Joseph Fields's great-grandparents had been slaves in South Carolina; here, despite all odds, his family had been farming for more than a century. Joseph, now in his seventies, guided me from row to row, showing off sweet onions and Carolina Gold rice and a variety of Muscadet only grown in the South. As he handed me my first taste of fresh green peanuts, his countenance held the joy he described of being on the land as a young boy—back when, he said, he would pick tomatoes that sold for ten cents a bucket.

He was connected to his Gullah Geechee culture, he was connected to this soil, but he was also connected to the economic reality of which Bonita had earlier spoken. "Farming: You gotta love it to wanna do it. It's a job," Joseph said as a butterfly flew by. In his lilting Gullah accent, he gave me the rundown from planting to harvesting to selling, plus the less thrilling aspects, like when hurricanes destroy your field, keeping records, and dealing with crop insurance.

"It's a lot of work," I said.

"It's a lot of work," he said. "It's a business. It's a business."

Business acumen is why Joseph, after hearing more customers ask for chemical-free produce, turned his farm to organic agriculture in 2008, becoming the first certified organic farmer in the Charleston area. And it was why he once made the eight-hour round-trip trek three times a week to Asheville, North Carolina: His market is diversified, and there his vegetables fetched a high price.

Nikki Seibert Kelley elaborated earlier how Joseph, as a successful farmer with a half century of experience, has a dizzying skill set few new farmers have. "You're expected to be a business person, a financial person, a crop planter, an environmentalist, a soil steward, a service expert . . . and then you also have to be savvy enough just to watch the other markets changing. So you have to be, like, a vegetable stockbroker," she said. This is why USDA defines a beginning farmer as one who's been farming for less than ten years—the duration of med school plus a residency in neurosurgery, or two PhDs. The learning curve of farming (along with the obstacle of finding affordable land) is why farming incubator programs are relegated to about an acre, why so many young farmers I met are creating value-added products (like cheese) to coax a living from a smidgen of land, and why some small farmers I followed failed (and failed to call me back) during the writing of this book. "There are very few people that you could sell or give a sixty-acre vegetable farm to," Nikki added. They don't have the chops for it.

And yet everyone who romanticizes living off the land thinks that they can start a farm, said Sarah Mock, the agricultural reporter who wrote *Farm (and Other F Words)*. When we spoke, she had just attended a conference where a man said he was pursuing farming after attending Outstanding in the Field, an outdoor table-to-farm dining event. She pushed back with what I will forever refer to as "Sarah's plumbing example": "OK, if you went to an award dinner for a plumbing association and got to go to a really cool plumbed facility and celebrate the coolest accomplishments of a group of plumbers, would your conclusion at the end of that night be: I can definitely be a plumber? And I should just start now?"

There may come a day, sooner than most of us realize, when another pandemic strikes or the ocean swallows Charleston whole, or industrial civilization unravels in any number of imagined ways, keeping us awake at night. We all should regain the skills to grow our own food and boost the food security of our communities. We also should develop alternative economies and collective farming and land ownership strategies by any of the innovative means outlined in this book.

But in the meantime, commercial farming happens within the context of capitalism. The latest research from Project Drawdown estimates we could remove up to twenty-three gigatons of excess carbon from the atmosphere by 2050 if we escalated regenerative annual cropping—*which means regenerative agriculture may have the potential to reverse climate change*—but not if sustainable farms can't actually succeed. Meanwhile, industrial agriculture is adopting regenerative-ish practices, like no-till in conjunction with glyphosate (see chapter 5) and automated plant factories (see chapter 3), but that's not the kind of cultural transformation we should be interested in.

Joseph told me he hoped to pass on his farm to his grandson, who is now nineteen. Joseph knows the obstacles to farming are monumental: rising seas, increasing hurricanes, encroaching development eating up what farmland is left. But he would never sell his land, he said.

We strolled toward his house to meet his wife, Helen. Joseph gathered up a handful of felled pecans and handed them to me. They now reside in the desk where I've been writing this book. (I plan to eat them when I turn in the manuscript.) He asked about my life in Los Angeles, and heartfelt presence that he has, I somehow let down my guard enough to mention that I went to school for classical voice.

"She sings opera!" he called out to Helen, who was sitting in the packing shed.

"OK, hit a high note for me," she requested, and I obliged with a high B. (I was jet-lagged; otherwise, I swear I would have gone for the C.)

"Are *you* a singer?" I asked.

"Mm-hmm," she said, then launched into a stirring mezzo-soprano.

Shackled by a heavy burden,
'Neath a load of guilt and shame.
Then the hand of Jesus touched me,
And now I am no longer the same.

He touched me, Oh He touched me,
And oh the joy that floods my soul!
Something happened and now I know,
He touched me and made me whole.

"I can't hit those high notes anymore," she smiled. "But I can do the best I can with what I've got left."

EIGHT

Grass Nomads

"Our task is to make trouble, to stir up potent
response to devastating events, as well as to
settle troubled waters and rebuild quiet places."

—Donna J. Haraway, *Staying with the Trouble*

ARIEL GREENWOOD'S STORY WAS STARTING OUT LIKE A TALE I would dub *Goldilocks in the Garden.* From her teen years through her early twenties, she tried out the life of a vegetable farmer—growing produce for community gardens, a private homestead, even a multibillion-dollar tech company.

"And that felt too small," she said.

I thought Ariel was speaking metaphorically. When we first met in 2020, I gleaned she is a young woman of deep thoughts and lofty, sometimes labyrinthine ideals. But it turned out Ariel felt as physically mismatched as Goldilocks. "I'm a very tall person, about five foot eleven. So me handling, you know, thinning carrot seedlings just felt ridiculous!"

Ariel, thirty-three, is a cattle grazier, painter, and writer. Throughout our years of scattered calls and correspondence, it never occurred to me that she could have been so statuesque. She's described herself as "skinny," but in photos of her riding horseback or roping sick cattle (to remove them from the herd and treat them) or gazing across grassland vistas—her blond hair always tucked into a big straw hat—she appeared to be a petite, graceful powerhouse.

The optical illusion was due to scale. Ariel and her husband, Sam Ryerson, forty-one, a rancher,[1] operate daily on a scope more

---

1 Rancher and grazier are used interchangeably, but Ariel elucidated the difference: She considers a grazier someone who manages grazing animals, whereas "rancher" has ownership connotations—either owning a ranch or owning cattle. "I usually call myself a land and livestock manager—or grazier, but I get tired of explaining terms these days," she wrote in a text. Technically, Sam is a rancher and a grazier.

monumental than most of us will ever experience. It was June, and Ariel was talking to me from the porch of her adobe house on a roughly 120,000-acre ranch outside the village of Wagon Mound (a post office and two gas stations) in northeastern New Mexico, where they would work through at least that winter. Sam is a partner in Triangle P, a five-person-owned cattle company that leases the land and owns the cows; Ariel and Sam co-own and operate Grass Nomads LLC, an ecologically focused company that grazes the cows and manages rangeland across the semi-arid West. The two handle hundreds of half-ton animals across a landscape extending from the high-elevation grassland of the Southern High Plains to the ponderosa pine forest of the Sangre de Cristo foothills, the southernmost subrange of the Rocky Mountains. It's an altitude span of roughly 6,400 to 9,200 feet.

The ranch's topography is as majestic as its vistas, from volcaniclastic sandstone and old lava flow to live creeks and streams that ebb and flow with the rainfall. There are thousands of elk and antelope that traverse the property. The climate, too, is equally impressive: I booked my flight to visit just before a winter of subzero temperatures that saw Ariel and Sam breaking thick ice for days on end. But then my kids came down with RSV and I had to reschedule. Now we were catching up by phone, and Ariel and Sam were in an early summer monsoon season. That morning, Ariel—on horseback—had raced a hail cloud back to cover her little container garden, previously smashed in a storm. (She hasn't entirely given up gardening. "This is my final attempt to get something to grow.")

Even considering the standard seasonal change they experience in New Mexico, Ariel and Sam had navigated a far more humbling environment in the past few years due to the colossal force recalibrating everything these days: climate change. Despite recent rains, the Southwest was in the midst of a megadrought drier than any in the past 1,200 years. The year prior, it fueled wildfires that nearly consumed the ranch—the largest wildfires in New Mexico history.[2] Not long after, at the ranches Ariel and Sam work in southern Montana, they found themselves in the midst of the historic floods that shut down Yellowstone National Park. Friends of theirs nearly lost their home, and other houses upstream were sucked away.

"But that's the nature of working in really big places—really big things can happen," Ariel said. "We were lucky."

Ariel majored in psychology in college, so I asked for her take on the psychology of someone who would tackle this occupation. "That's a great question," she laughed. I pictured her staring at the former fiery hellscape now turned heaven on earth, as she described it, with gently rolling hills covered in green.

"I'm not outdoorsy in the way that a lot of people talk about, like hiking, skiing, and so on. It's more: That's where I choose to live . . . that's my comfort zone." But there was one common thread she noticed between herself and other friends who hadn't grown up in, but had chosen, this intrepid way of life.

"I think every person has a kind of physical scale that they're comfortable working in. And for me, it's a pretty large scale," she said. "I just feel comfortable in big places around large animals and working in jobs that have a lot of big questions around them."

Ariel, after all, was no helpless Goldilocks heroine. For her, Sam, and a new movement of first-generation ranchers and graziers, the biggest chair is the one that fits.

---

2 One of the megafires was started by a Forest Service–prescribed burn that sparked out of control. Needless to say, this didn't go over well in the already politically and socially charged state. "People are still very angry," Ariel said.

I FIRST REACHED OUT to Ariel three years ago because I was mesmerized by her way of life. Working on the farm in Oregon, I had one day been too nervous to try out a chainsaw (or was it a Weedwacker?). And yet there was Ariel on Instagram, starting colts and living out of a trailer.

After our first call, she sent me an email: "I think you could write your whole book about adventuresome people looking to livestock as a means of forging a regenerative path on the planet!" (Probably; but that would have necessitated a journalist who could ride a horse.) It was then that she introduced me to her circle of ranchers—many young women graziers like herself. They include her close friend Brittany Cole Bush, who worked as an "urban shepherdess" outside Oakland, California, and now runs her grazing business, Shepherdess Land & Livestock Co., from Ojai. There, she employs goats and sheep to chomp the brush that fuels wildfires. Another colleague, Ruthie King of Shear Mendocino, is a sheep shearer, farm educator, and volunteer wildland firefighter who commenced her unconventional path with a degree in architecture and sustainable development from New York's Columbia University. Several others Ariel mentioned were channeling "pro-environment, pro-social values" into grazing, vegetable farming, and land tending in California.

These ranchers, it appeared, had ideologically focused callings. However, the course Ariel and Sam were on in 2020 was murkier. "We're sort of like the goat graziers of the commodity cattle industry," she explained. Ariel learned how to manage livestock in California first on a nature preserve and then for a private ranch, raising grass-fed beef. "My whole context for working with cattle has been working with the demand for beef and meat to try to improve the land on which they're grazed and managed. It's an awkward relationship, and one that's not designed to improve the land," she conceded.

Back then, Ariel and Sam were anomalies in the large-scale ranching world, which is one reason they found each other. (There was a cowboy/cowgirl romance, too, but I'll get to that later.) "We've come into this work because it is a way for us to express our values, but a lot of our values are not born of this work, if that makes sense," she told me.

In other words, they not only worked in the big business of beef cattle but in the realm of cognitive dissonance: preservationists in a polluting industry that is the target of environmentalists and vegans everywhere. It's not what one would expect of Ariel, a former small-scale organic vegetable farmer who minored in agroecology in college. Yet while agroecology may sound like a hippie course of study, it is, fascinatingly, a bridge between the technological and the natural; a discipline that aims to bring ecological principles into established food systems. This, it turned out, would be Ariel and Sam's long game.

Still, Ariel encapsulated their conflicted situation like any typical cattlewoman—by quoting ecofeminist cyborg scholar Donna Haraway. "She talks about going where the trouble is. So that's why I stay in ranching."

AS A CHILD, ARIEL had ample space to contemplate life's convoluted questions. She grew up largely unschooled in the rural environs of Vance County, North Carolina.[3] She described her parents as

---

3 Unschooling is just what it sounds like: no formal schooling. The self-directed learning approach falls under the umbrella of homeschooling, which is legal in all fifty states. Since this is a book about farming, it's worth noting that widespread formalized schooling didn't exist until America's move to a post-agrarian society, when new generations had to be "educated" to enter a specialized and stratified industrial (now, technological) workforce. Forty-nine percent of Americans were still farmers in the year 1880. The movement for compulsory public education didn't begin until the 1920s.

"pretty religious," though "also educated": Her father was alternately a preacher and minister or insurance salesman; her mother did secretarial work and later became a property manager. Ariel is the youngest of three children and, educationally speaking, said she was "the product of benign neglect": "I had a lot of land to explore, some pet horses and chickens . . . a lot of latitude to do my own thing and a lot of autonomy." Clearly her learning wasn't lacking: She enrolled in North Carolina's Vance-Granville Community College at the precocious age of sixteen, right around the time her family moved to a bigger house on a smaller lot.

Missing the swath of nature in which she once wandered, Ariel joined the college's ecology club, where a field trip took her to Melvin's Gardens, a small organic perennial and herb farm not far from her home. She started volunteering there, "just to have something to do outside," and was hired at the age of seventeen. She later tried her hand at those aforementioned "too-small" vegetable farms and advanced her studies at North Carolina State University. It was there that she began learning about grassland ecosystems, those millions-year-old landscapes that evolved in concert with a web of predator and ruminant megafauna—the ancestors to today's cows, goats, pigs, and sheep. On America's prairies, that regenerative relationship persisted for thousands of years until the late 1800s, with the near extinction of bison, as well as that of the Native people who relied on this mainstay of their food system.[4]

At the same time that she was learning about grassland evolution, Ariel started meeting people who were "holistic managers" of livestock. The holistic management approach originated with Allan Savory, a Zimbabwean farmer who has since garnered controversy

---

4 Notably, the hunting of bison, and subsequently their near extinction, was a pointed tack of genocide by the US Army, which doled out free ammunition to hunters. In the words of one colonel: "Kill every buffalo you can! Every buffalo dead is an Indian gone."

for his unsupported claim that grass-fed beef alone can reverse desertification (even where there should naturally be deserts) and global warming. The term *holistic management* is now a registered trademark of Holistic Management International, an organization that teaches ranchers and farmers how to incorporate regenerative agricultural practices into viable businesses that meet the demand for local food.

This method offers a dose of pragmatism not only to Savory but, at the other end of the spectrum, to dystopian environmentalists like George Monbiot, who would like to see wild spaces cordoned off from humans and those same humans subsist on lab-grown meat substitutes produced in factories. (Never mind that Monbiot's approach would require vast energy and resources inputted from those wild places.) The reality is that livestock grazing encompasses 25 percent of all land on the planet, and humanity's relationship with herding is an ancient one—one that's been sustainable for many cultures, from Mongolian nomads to the Maasai.

That being said, worldwide livestock grazing—which has degraded 20 to 35 percent of the earth's permanent pastureland—is far from Maasai-esque. Yet holistic management presents a considerable chance for renewal. It includes holistic grazing—a method of grazing livestock that stewards land by controlling the frequency and density of animals on pasture. According to Project Drawdown, this type of managed grazing could sequester some twenty-one gigatons of carbon dioxide by the year 2050 and offers an increasingly viable alternative to the devastating ecological impact of feedlot farming.[5]

"To make a long story short, my first exposure to livestock was

---

[5] Concentrated animal feeding operations (CAFOs), better known as factory farms, have led to the rise of antibiotic-resistant superbugs and produce up to twenty times more waste (i.e., feces) than American humans do each year. Regardless of how you feel about meat, there is no doubt that raising cattle on pasture is healthier for the environment, people, and the animals themselves.

through this ecological lens," Ariel said. "And so that's what attracted me to it, whereas a lot of people [in ranching] start with the production lens and then learn the ecological stuff out of necessity."

So Ariel moved to California to learn the art and science of cattle ranching—where she "fumbled [her] way into a practically unpaid, kind of ridiculous internship" on the aforementioned nature preserve, then took the job on a grass-fed ranch in western Sonoma County. There, she found herself attracted to another regenerative relationship—or, rather, a rancher named Sam, introduced to Ariel via email for a potential job with her ranch. He didn't take it, but they stayed in touch professionally. Holistic graziers in the commodity cattle world were then few and far between.

"Eventually, we were both single at the same time," she laughed. "He was managing some yearlings just outside of Wagon Mound . . . and living in a wall tent and didn't have good cell reception. So he started writing me letters. And, you know, once you start receiving letters from a cowboy in a wall tent, you kind of let your guard down a little bit."

For six months, they visited each other at their respective ranches. Then she quit her job and moved into the wall tent—by then on a ranch on an Apache reservation. When I first talked to Ariel in December 2020, that lease had ended, they were about a year into their current New Mexico gig, and they had upgraded to a camper. That first winter on the property had been rough, she said: "cold and dry; the place was really droughted out and grazed pretty hard." When we reconnected in June 2023, she didn't take her relative comfort for granted. "Now that I'm sitting on the porch of our house, it feels pretty bougie," she chuckled.

Like Ariel, Sam is a first-generation rancher. He grew up in Cambridge, Massachusetts, and went to Yale—an intelligent self-starter type who's skilled with both machines and animals. (That's according to Ariel. Also, "He is really shy—especially digitally—and doesn't enjoy talking about himself or his story much," she wrote, hence him declining our interview.) But unlike Ariel, whose training was essentially trial by fire, Sam was one of the first apprentices with the New Agrarian Program (NAP), run by the New Mexico–based nonprofit Quivira Coalition. It offers eight-month apprenticeships in regenerative ranching.

Since 2008, NAP has graduated more than a hundred apprentices and now partners with established farms and ranches across Montana, Colorado, California, and New Mexico to train young folks in operations from organic beef to grass-fed lamb to holistic dairy and even integrated grain farming. And since Sam's days, the program has scaled considerably: NAP now partners with twenty-five to thirty ranches each year, each one taking on one to two greenhorns. There will undoubtedly be more regenerative ranches for aspiring ranchers to partner with in the years to come: Since our first conversation back in 2020, Ariel said holistic management practices have rapidly infiltrated conventional ranching. "There's been a little bit of diffusion of those principles, which some people think is not so great, but I think it's all good," she said.

Even though many NAP apprentices come from nonagricultural backgrounds, more than 80 percent of graduates are either working in or toward an agricultural career, said Quivira Coalition Executive Director Sarah Wentzel-Fisher. This is hopeful, considering the imperiled state of American ranching: The average age of a US rancher is fifty-eight, and millions of acres of private rural land are at risk as these ranchers head toward retirement. This is especially true in states like Texas, where 95 percent of land is privately owned. All that ranchland could one day become subdivisions or Monbiot fake meat mills.

Public land is vulnerable too: Many US graziers lease their land from the Bureau of Land Management, some 155 million rural acres. After all, who can afford to buy a 120,000-acre ranch these days? "Those public land leaseholders are aging out also," Sarah said. Meanwhile, by its own benchmark, the Bureau of Land Management's rangelands are gravely damaged by overgrazing. "Public land agencies aren't stepping up and saying, 'How are we recruiting, supporting, and putting people in the right places where livestock is really needed to maintain our shared public landscape in a positive way?'"

But NAP's success rate is encouraging, considering the learning curve of ranching. From my conversations with Ariel, I would place her job description in the realm of *ecologist-hydrologist-geneticist-animal behaviorist-veterinarian-master horsewoman able to ride over dodgy terrain up to twenty miles a day*. To wit: When I asked if she recently noticed anything on the landscape that indicated how their grazing practices are fostering biodiversity, she enthused for five minutes about the rapid increase of western wheatgrass, a cool season native grass species with high protein that fills in bare ground and improves the bioregion's hydrological cycle. Then she delved into its implications for the nutritional health of the cattle as well as the process of genetically tailoring the animals to the resources of the place. (I had been expecting her to say she noticed more birds.)

So what was a typical day in June like for Ariel? Well, that week she would move hundreds of cows on and off pasture, sort animals, brand animals, attend to their health, disperse salt or mineral supplementation if needed,[6] check fences, survey water, turn pumps on and off, and monitor rangeland, plus plan upcoming stream restoration and infrastructure improvement projects with the Natural Resources Conservation Service.

And this was summer, their slower time, plus Ariel was working "part time" six or seven hours a day. She hadn't responded to

---

6 Because Ariel and Sam produce cows adapted to the environment, minimal supplementation is needed.

my emails earlier that spring and now I knew why: She was sixteen weeks pregnant with their first child.

"I'm feeling better now that I'm out of the first trimester," she said. "My energy level is definitely not what it was, which is pretty frustrating, especially [because] we're having a really great season here . . . everything's turning green and water's flowing. There's animals everywhere."

Her reluctantly "slower" pace, though, was a chance to lean into another personality trait that drew her to this larger-than-life way of life: "I think I'm pretty ADHD," she said. "A lot of people are who work in this kind of thing. It's our relationship with dopamine, and that has to do with exploring curiosities more so than achieving things." That day, she had spied the first baby antelope and baby elk of the season. She found a horny toad and spotted springs that hadn't been running in a while. "That, to me, is a lot of the bread and butter," Ariel said.

There would be ever-present puzzles to solve, however. As she earlier described, "Ranching is a really interesting mash-up of very intellectual rigorous work and decision-making with . . . a lot of physical finesse and elegance, things that are associated with roping and riding and starting colts and working with animals that weigh more than you do but have their own experiences and psychologies and emotions." She added, "There's not really that much brute force in ranching. That's kind of a trope."

Then, too, there would be a child to raise—an adventure in and of itself on a ranch. "It's probably one of the best ways to grow up, but as a parent, especially as a mother who's also passionate about the work, I can imagine it'll be pretty challenging," she admitted. But in time, after having the baby, she would literally get back in the saddle again.

"There's not as many of us as there are small-scale farmers and vegetable growers, probably because the barriers to entry are higher," Ariel once told me.

Sarah at NAP had concurred: "There's nothing easy about entering [regenerative ranching]. But I also think that people find it extremely rewarding, very tangible, and real and meaningful in ways that a lot of young people are yearning for right now."

NINE

Rooted Northwest

"We have removed what I think are
the two most important things for
a vital life: connection to nature
and connection to each other."

—Steve Nygren

FIVE YEARS AGO, FOR TWO GLORIOUS MONTHS, MY FAMILY AND I lived in Serenbe, a "utopian wellness" community, as it's been widely described, nestled into 40,000 acres of protected forest and rural land forty minutes south of Atlanta. My husband, who is a screenwriter, was filming a TV show in the area. We originally thought he'd be traveling solo from Los Angeles to Atlanta, but then coincidentally (or as fate would have it), the April week he heard about the show's filming location we found out the lease on our house was being terminated. And so, in a foreshadowing of our later pandemic adventures, we sold the bulk of our furniture, put the remainder in two orange pods, pulled the kids out of school early, and packed a duffel for paradise.

Our daughters were then seven and five, and Serenbe—an amalgam of *serenity* and *be*—has the makings of one of the most idyllic places in modern-day America to raise one's kids. The community of over 650 residents in Georgia's Chattahoochee Hills, started in 2004 by restaurateur-turned-developer Steve Nygren, is what's known as an *agrihood*, the now fast-growing residential development trend that forgoes the golf course or tennis complex to situate houses amid a working farm. According to the Urban Land Institute, there are now more than 200 agrihoods across twenty-eight US states, although Serenbe remains the most visionary version I have seen: Nygren laid out the thousand-acre community like the English countryside, with fifteen miles of wooded trails connecting three compact, walkable hamlets, each with impeccably designed (think *Architectural Digest*–level) homes and mixed-use dwellings. Cars are few and far between; the whole property is traversable by foot, bike, or golf cart. There are "secret" treehouses for capering in the forest. There are blueberry bushes lining the streets for anytime face-stuffing or muffin-baking. And at the center of the community is a twenty-five-acre certified organic farm, where residents can procure 300 varieties of hyperlocal veggies, herbs, and flowers throughout the growing season (April to November) and even volunteer.[1]

Dreamy and privileged as this sounds, Nygren, whom I interviewed then for my podcast, doesn't favor the descriptor *utopia*, with its connotation of unattainability. He instead defines the community as guided by *biophilia*—the term popularized by the late biologist E. O. Wilson to describe our essential human need to live in concert with nature and its living creatures. Just in case you're not sure how our nature-disconnected, smartphone-addicted way of life has led to an epidemic of loneliness and illness, take a look at the death rates from suicide, alcohol, and drug overdoses, now at an all-time high. Seventy-four percent of US adults are classified as obese or overweight, and 20 percent of children and teens are obese—predisposing them to diabetes, heart disease, cancer, and an untimely death from all causes. The average American consumes a hundred pounds of sugar and sweeteners a year, and a typical piece of produce travels 1,500 miles from farm to plate. These truths are now *normal*, and yet they are not the human *norm*. Biophilic agrihoods are a fancy term and innovative means for restoring what we all not just *should*, but *need* to be doing every day: spending time outside, eating fresh food that grows nearby, and interacting with each other.

And undoubtedly, biophilia was the experience of our Serenbe stay. When my daughters weren't catching frogs in the forest or

---

1 The farm cultivates ten acres and is currently restoring soil on an additional fifteen.

spying horses at the stables, we were trekking across pastureland to nuzzle goats, sheep, and llamas in the "animal village," occasionally outsprinting thunderstorms on the way home. Later, in the community's summer camp, the girls canoed on the pond and rambled to waterfalls. Meanwhile, I went trail running in the woods, snacked on wild blackberries, and brainstormed book ideas. (My poor husband was working twelve- to fourteen-hour days during all of this.)

And then there were the daily opportunities for multigenerational social interaction, a boon to my kids' well-being and to parents like us who had been raising our children with scant family nearby: the grassy commons' in-ground trampoline that guaranteed a stream of playmates for my girls; the General Store, where owner Nadine Kzirian taught my then seven-year-old how to run the checkout line; and our impromptu evening trips to the inn for a scotch (lemonade for my daughters) with a lively patrician woman in her seventies.

But nothing made me feel more, um, biophilic than walking down a cobblestoned lane to the farm on CSA pickup day, when I could collect turnips, kale, zucchini, microgreens, and basil in a basket and head straight back to our kitchen. This is where the seed for this book was planted, although farming then, like the rest of Serenbe's luxury agrihood existence, appeared to me as aspirational fantasy. Only later, during my own farming internship and the research for this book, did it occur to me to ask: Who owns the land? How much are the farmers paid? Could they afford to live there?

It was in answering those questions that I recognized who wasn't vested in that visionary agricultural community of flora, fauna, and people: the farmers.

Serenbe (i.e., the Nygren family) owns the land. The farm manager may be well compensated, but it's a hired position, one that changed hands several times even before I arrived. The farm's interns receive a $1,000 monthly stipend and free housing, but I'm guessing when their internship is over they won't stay in a community where a one-bedroom condo lists for $500,000 and most houses are over $1 million.[2] Even Serenbe's farm animals, I realized, are superficial, reserved for a petting zoo while the farm grows only vegetables. This, despite the fact that livestock integration would not only build soil but also a more diversified local diet for residents who, when I was there, shopped for meat, dairy, and staples at chain supermarkets in the outside suburbs.

Of course, the romanticization of #farmlife is what agrihoods are selling. Not that this is a strictly bad thing, given the alternative they offer for a healthier way of life. And given an exploding world population and ceaseless development, agrihoods also offer a novel approach to land conservation—especially farm- and ranchland, which is disappearing in the United States at the rate of 175 acres every hour.

In fact, preserving land was the impetus for starting Serenbe: After Nygren sold his restaurant business in 1994, he and his family moved from Atlanta to their vacation farmhouse in this bucolic area, where, after witnessing a bulldozer clear-cut, he offered to buy land from neighbors who were ready to sell, to protect it from Atlanta's then burgeoning sprawl. (He wasn't expecting them all to take him up on his pitch and wound up with 900 acres—the land that became Serenbe.) Later, Nygren worked with 500 landowners in Chatta-hoochee Hills to change the zoning so that today, 70 percent of land is conserved as farm or forest, and just 30 percent can be developed for housing.[3]

---

2 Life in Serenbe doesn't come cheap, but it's relative: Our rent for a furnished two-bedroom apartment was $3,800 a month—about what we had been paying for a 1,100-square-foot house in Los Angeles, without the access to a farm and a 40,000-acre nature preserve.

3 Conventional zoning would have allowed construction on over 80 percent of the land.

Farmland preservation was also an aim for Kiawah River, a new 2,000-acre agrihood I visited on Johns Island back in Charleston, South Carolina—no small feat where, just across the river, habitat had been obliterated in the name of five public and two private golf courses.[4] Kiawah River instead devoted 100 acres to sustainable agriculture—including vegetable farming, goat dairying, chicken ranching, cattle grazing, and beekeeping. "At a time like COVID, when everybody was moving [to Charleston], any other developer would have come in and done everything to turn every piece of land over into a home," said farmer Missy Farkouh, who co-owns The Goatery at Kiawah River, which provides milk and cheese as well as agritourism offerings such as goat yoga.

But as with Serenbe, the Kiawah River agrihood—replete with twenty miles of waterfront, an on-site Auberge hotel, and $1 million to $6 million homes—is likely not within the reach of most farmers. Still, Missy, who lives thirty minutes away from Kiawah River in the suburb of Mount Pleasant, South Carolina, and made the switch to goat farming after "working [my] butt off for twenty-five years in corporate America," passionately described her community of farmers and their partnership with their developer as a win-win: The farmers gain a direct market for their products plus enviously affordable land access (Kiawah River charges them $1 yearly rent and takes no cut of their sales); Kiawah River's residents have access to a near-whole-diet farm share and a connection to the people who grow and raise their food.[5]

Yet when I toured Kiawah River, which is in a FEMA Special

---

4 One in four US golf courses was built as part of a housing development. Could you imagine the local food systems we could create if those were farms instead?

5 Whole-diet CSA programs are increasing in popularity. They allow eaters to trade the grocery store for a farm, procuring their food (vegetables, fruit, meat, eggs, grains, dairy, etc.) from a single source.

Flood Hazard Area, and zipped in a golf cart past a resident waltz-ing with a glass of white wine amid gargantuan homes on elevated foundations, it felt a bit like farming was being used to market the *Titanic*.

"It is so easy to sell real estate here," said Chris Drury, Kiawah River's broker in charge, as he golf-carted me to my car at the end of my tour. "Don't tell anybody else I said that."

"Well, my tape recorder is still running," I said.

"Oh, OK," he conceded, before rhapsodizing a wee bit more about how the idyllic place sells itself. On cue, a breathtaking doe leaped across my path before I drove away.

Agrihoods can be noble endeavors (or just for nobles), but they're also real estate strategies—in service of the developer, not the farmer. But a hot new take on the farming communes of yesteryear could bolster the back-to-the-land movement in a more equitable and enduring way.

⁂

I HAD JUST HOPPED on the phone with permaculture designer and educator Dave Boehnlein when he had to interrupt our call because a wasp was hovering in the room. Dave, forty-four, is a permaculture designer and educator who coauthored the primer *Practical Perma-culture*, so of course, when he rejoined the line, I had to annoyingly ask him about the permaculture approach to handling a stinging insect in the room.

"I got a flyswatter," he said, laughing. "I'm partially allergic. Catch and release was not the name of the game this morning."

Similarly, Dave described an adaptable approach as the cofounder and project manager of Rooted Northwest, a cohous-ing "agrivillage" he and (at the time) thirty-five members were in the midst of cocreating on 240 acres of community-owned forest, pasture, and cropland outside Arlington, Washington. (I realize I just used the prefix co- three times, but togetherness, as you'll see momentarily, is the gist of this undertaking.) "If we were to have a toast, our toast would be, 'May our mistakes be original,'" he said about the outset of the project. "We didn't have the hubris to think we weren't gonna screw things up in some way, shape, or form. But we certainly didn't want to screw things up in the same way that people have screwed things up over and over again."

Unlike the real estate paradigm of an agrihood like Serenbe or Kiawah River, Rooted Northwest is employing a more uncharted model, at least for most Americans: cohousing, a type of intentional community originated in Denmark where members collectively own and share land, facilities, and resources, and also cooperatively govern. Rooted Northwest uses a governance system called sociocracy. Cohousing, however, is no hippie commune of the 1960s: Residents independently own the homes on that shared property and have the privacy of traditional features like one's own kitchen and a fenced yard.

In Rooted Northwest, up to seventy homes will be clustered in a five- to ten-acre neighborhood, with a common house and kitchen that might host community dinners three times a week. There also will be joint amenities, such as laundry and bike storage. But the most sizable shared space—both physically and conceptually—will be the farmland: more than 180 acres of prime soil in the lush western foothills of the Cascade Mountains, all devoted to regenerative agriculture.

Whereas the typical agrihood resident is seeking access to locally grown food but not a farming lifestyle, the envisioned Rooted Northwest inhabitant is actually a farmer. Dave explained that not all community members may farm, but everyone will support farming—whether that's lending a hand with labor or simply welcoming the morning crow of roosters. As with the private cohousing residences, each farmer will have autonomy: the shared farmland will host a mosaic of microenterprises that can collaborate or not.

Dave wasn't taking the path to success lightly. Intentional communities, if you remember from the isolated roadkill-eating Wild Roots community in Appalachia, don't always manifest the way they were intended. Dave cited the oft-repeated statistic that nine out of ten intentional communities fail. "We weren't interested in being a part of that because we figured there's probably some patterns there," he said. I thought he might delve into examples of juicy personality-fueled conflicts, but he listed more prosaic (though just as pivotal) concerns for a project outside conventional real estate development: "not [acquiring] your land properly; not creating a [legal] entity to hold your land; and not setting up accurate expectations about what you're doing from the get-go."

Dave was evidently a visionary with some serious execution skills. In the early 2000s, he lived pretty radically (a one-room cabin with no electricity and a community kitchen) for seven years with three families and a dozen interns at Bullock's Permaculture Homestead on Orcas Island, Washington, where as education director he "wore all the hats no one else wanted to wear"—editing the newsletter, running the website, and doing general admin. He also helms a company, Terra Phoenix Design (with the Bullock homestead's Doug Bullock and industrial designer Paul Kearsley), which designs permaculture master plans in environs as ecologically diverse as the Greek Islands and the Peruvian Amazon.

In 2011, Dave made the move from island homestead life to Seattle to be with his now wife, Yuko Miki, an artist/illustrator who was raised in Japan and grew up in a farming family. He wrote his book, then was recruited by a young couple who hoped to raise their kids on a farm and consulted him for design advice about a different permaculture-inspired property. But a year and a half in, the couple realized it wasn't what they hoped for and approached Dave and

Yuko about co-launching something, the four of them. "I said, 'Well, you know, I don't have anything else to do for the next ten years, so why not?'" he quipped. They set a vision for Rooted Northwest and got the ball rolling, setting out to find inspiration and land.

The two couples were far from farm commune–seeking society dropouts: Argentina natives Eduardo Jezierski and Elina Quiroga are, respectively, a former CEO/CTO of a multinational nonprofit who has a background in physics and AI/machine learning; and a vascular trauma surgeon. Needless to say, this dynamic group did their due diligence in laying the groundwork for the project.

First off, they spent a year and a half visiting all the intentional communities and farming projects they could find, ultimately landing on a mash-up of three of the most inspiring: for the housing, New York's thirty-year-strong EcoVillage at Ithaca, where the passive solar residences in the FROG (First Residents Group) neighborhood just "felt so good"; for the farm, Our Table Cooperative outside Portland, Oregon, "the first multi-stakeholder [farmers, producers, consumers] agricultural cooperative out there"; and, because they eventually want to teach others, California's Occidental Arts & Ecology Center. "It's both an intentional community and an education center," Dave said. "They've actually found a way to weave those things together so that people living there don't feel like they're in a fishbowl."

Then they found land—what Dave calls "the unicorn": an old dairy farm locally stewarded for a century, with fertile soil and at risk of development because the owner was moving on in years and getting ready to retire. The 240-acre farm was actually zoned as a residential five-acre minimum zone, which meant—in the hands of a conventional developer—it would be covered in five-acre lots with houses. "So there would be no more farm," Dave said.

By that point, their group had started to draw other families into the project, allowing them to buy the land. Now, Dave said, they're pushing through the permitting process with Washington's Snohomish County. Saving farmland by creating a regenerative farming community hadn't been a tough argument to make ("I don't care what your politics are; who doesn't want to see ag land preserved?" Dave challenged), but as Serenbe's Nygren first discovered, the obstacle has been the building code itself, which is not designed to condense rural housing to leave swaths of land intact.

Luckily, county officials have been "super supportive." Dave, it turns out, was preaching to the choir. "They're really into agroforestry [integrating trees and shrubs with crops], silvopasture [adding grazing livestock to those integrated trees], biochar [making a carbon-rich material from biomass to boost the fertility of soil], minimal and no tillage agriculture [no digging; see chapter 5]—all the regenerative ag buzzwords are the things they're trying to get people to do."

The second piece of the puzzle will be Rooted Northwest's housing: creating the ideal situation so that farmers can once again live where they farm. Living where one farms may have been the historical norm, but soaring land prices in this day and age have relegated this privilege to the wealthy (wealth being in cash or inherited land).[6] "The idea that [a farmer] doesn't have to commute thirty minutes or an hour to get to their farm every day while they live somewhere else is a really appealing part [of Rooted Northwest]," Dave said. The markets for farmers, too, are enviably close: Downtown Arlington is biking distance, and Seattle and Bellingham, a college town with a thriving craft beer and foodie scene, are an hour by car. Not

---

6 Many farmers are cash poor, hence the reality that 56 percent of American farmers have a primary off-farm job. But despite the stereotype, most farmers aren't actually poor since land is a lucrative and increasingly desirable asset. Once we acknowledge the vast generational and corporate wealth that underpins American farming, we can advance small, sustainable farming models that might actually succeed. See Sarah Mock's *Farm (and Other F Words): The Rise and Fall of the Small Family Farm*.

surprisingly, small, sustainable farms in the region are flourishing—including Sonder Farmstead, the half-acre vegetable farm and herbal apothecary started by my former Mahonia Gardens coworker.

Houses at Rooted Northwest, which will attune sustainability with affordability, are anticipated to cost (drumroll, please) $600,000 to $1.2 million. Yes, these are Serenbe prices. However, the community is working on subsidizing strategies for farming residents. Yet the business advantage of the cohousing model for farmers, especially start-up ones, is that they are beholden to neither bank nor builder for their farmland, since it is collectively owned. The opportunities for collaboration are numerous, and the ultimate risk can be shared. "When the small family farm doesn't work out, the bank ends up owning it. And the farmer feels like a personal failure because they've bought into this individualist crap that doesn't really make sense, especially when . . . farming is so high risk and the margins are so small. It's less of a career path and more of an addiction," Dave said. "In our model, if one person fails, that's OK because hopefully there will be twenty different enterprises out here happening and multiple farmers involved."

Personally, Dave would love to see those regenerative farming ventures evolve into a co-op, but that will ultimately be determined by the people who come onboard.[7] When I spoke to Dave, the land was still a hayfield surrounded by hedgerows and trees, but the first vegetable farmer was about to come in. The group had just built a couple of caterpillar tunnels for him in anticipation of the "real wet" Pacific Northwest winter. "Bring your muck boots," he laughed when I pitched him a visit. In partnership with the local conservation district, which received a grant for experimental agroforestry in wet ground, they were planting three acres of basket willows (usable

---

7 Some successful (not all regenerative) farming co-ops: Organic Valley, the Tillamook County Creamery Association, Ocean Spray, and Sunkist.

for weaving baskets and natural building) and aronia (chokeberries; they're edible). There was a beekeeper who was readying some hives; other members were organizing a berry-growing collective.[8] Down the line, Dave could see himself planting hazelnuts and chestnuts. Future livestock farmers, he said, could run their pigs through the food forest to glean the fallen nuts. The whole community was hoping for a move-in date within the next three to four years.

But the one steadfast part of the dream, he said, would be the families with kids. Depending on who showed up, there could be a homeschool or programs for sustainability and nature awareness. Dave didn't have children, but as an educator (and, of course, a human) he couldn't wait to see them: "little packs of kids running off to the forest and building forts and then running out in the field."

The beauty of the collaborative approach is that no developer can dictate how you should farm or raise your children or what your biophilic home might look like. It's a choose-your-own adventure, a chance for a culture—and a food culture—to once again emerge from a tended landscape, just like it did over the course of human history.

---

8 I would be remiss not to mention a less privileged berry collective powerhouse thirty minutes north in Burlington, Washington: Familias Unidas por la Justicia, the first Indigenous (Triqui and Mixteco) farmworkers union in the United States.

TEN

Jewish Farmer Network

"When you are cut off from your past, that
past takes a stronger hold on your emotions."

—Claudia Roden

ALEXANDRA ("ALEX") ROSENBERG-RIGUTTO, A FARMER AND FOOD educator, came from a mac-and-cheese kind of family. Today, she is the farm director at Farber Farm, a one-acre diversified demonstration farm at Tamarack Camps in Michigan, a summer camp hosting more than a thousand kids from Detroit's Jewish community. But her childhood in a predominantly Jewish area of metro Detroit, as she recounted it now at twenty-nine, could have been that of any agriculturally disconnected suburban kid of the nineties/aughts.

Her family had flowers and ornamental plants to make their house look nice, but they didn't have a garden. Her parents are middle-class professionals: Alex's mom was a lawyer who now owns a consignment clothing store; her dad is a doctor, a family practitioner, serving primarily low-income folks on Medicaid and Medicare—people trying to meet their basic needs. "It's definitely from a paradigm of conventional, drug-focused medicine, which is ... helpful and relevant for people, but it's not preventative care," Alex said. As such, healthy eating isn't a focus of her dad's practice and wasn't a topic of conversation at home. "Talking about where food came from wasn't something that was an emphasis with us at meals or anything. There was no emphasis on organic or even vegetables," she said, followed by the parental mantra of our frenzied times: "You know, my mom did her best. Both of my parents were working a lot."

Alex's school life started out just as middling—she never felt comfortable in a conventional learning environment—then morphed into tough. She had a hard time making friends. From a very young age she was extraordinarily tall. "I'm five foot eleven now, and I feel like I've been five foot eleven since I was ten. That was always really hard!" I couldn't tell on the phone if her laugh was tinged with tears, but I saw where Alex's adolescent self was heading once she added: "I was chunky." (This was a generation before body positivity and a decade before the CDC released its first federal definition of bullying.) "I was such a target for [both] boys and girls to make fun of," she said. "I wasn't doing well in school, [though] I was really creative—I wanted to do art; I was always reading. But I was just kind of an odd kid."

So to avoid being a target, Alex took the path of many misunderstood artistic kids confined by suburbia: First, she played up her natural goofiness and became the class troublemaker; then she found an avenue to acceptance by a group of older kids. "I started using drugs at a really young age, and I became a very quick, um, drug addict," she said. Alex narrated this with the distance of being a together adult—an open book who had told this story many times. "It's something I'm integrated with and accept about myself because it brought me here and I'm really happy now." Still, she warned me with a nervous chuckle, "This is where it starts getting into a bit of *whoa* territory!"

Though I'm personally acquainted with traumas that drive difficult adolescences, I was not anticipating Alex's tale to take such a dark turn. Her job at Farber Farm is, after all, centered around childhood joy. Within five minutes of speaking with her, I surmised she is one of the most gratified people I have ever met, farmer or otherwise. And honestly, I was simply surprised: I had prepped for our interview by delving into Alex's farmsteading adventures on Instagram, seen her radiant smile and trim figure, and envisaged we'd be chatting about her black currant fruit leather, backyard-gathered morels, or how she deglazes a pan with medicinal broth. You'd think by this point I'd be wise to the empath types drawn to a back-to-the-land life (especially since I'm one of them), but apparently not.

Long story short: Alex was pulled from her freshman year of high school and put in inpatient rehab. She relapsed; there were suicide attempts and more hospitalizations. Al, a nice Jewish girl from the Detroit suburbs, became a high school dropout in tenth grade.

"The area that I was from, the environment that I grew up in—there was nothing for me to find and attach myself to that was generative, positive, or felt good," she said. "I just never felt at home at my home."

Alex's parents, fearing their child might die, decided to try removing Alex from that negative environment. So when she was seventeen, they sent her to live with her aunt and uncle in Phoenix, Arizona. But while she may have crossed the country, the setting seemed the same. "I didn't go to a hippie aunt's house who was eating congee and going to the farmers' market," Alex laughed. "I went from one suburb to another." In Phoenix, she found a new crowd of enablers and spiraled into unwellness. "I was eating so much fast food. I was so overweight. I was in pain. I was miserable."

Alex's aunt gave it to her straight. "She said, 'You look horrible!' which was not very nice," Alex laughed at the recollection, then imitated her aunt's suggestion that followed: *Oh my gosh, we live in Arizona. It's beautiful weather most of the year. Why don't you start walking or running or hiking or something?* So Alex obliged, first to get away from her aunt's scrutiny, then willingly and regularly driving to the edge of the desert, where she loved to just walk. "I felt so peaceful," she said. For the first time, she felt connected to a landscape. "Being in an arid, desert environment leaves a lot of space for thought. And I just started thinking and thinking and thinking about my life and where I wanted to go. Did I want to be [sick] forever? Do I want to live a long life? Do I want to live a happy, healthy life? I was really able to find this spaciousness in those moments that

[I realized]: *Yes, I did. And where could that start?* And that started with being outside more. And maybe I needed to look at what I was eating."

She added: "I think that right now with some of the body positivity stuff, there's a lot of pushback against wanting to lose weight. But it's OK to want to change your body in either direction. And it's OK to be where you're at in either direction. And I wanted to lose weight."

One day, Alex came down with food poisoning from a fast-food meal and connected the dots: *The food I'm eating is hurting me. But what if I did the opposite?* At twenty years old, she was on "a lot of medication"—including anxiety and depression medications and an anti-inflammatory drug for a skin condition. "The more I started researching food, and [reading] books, and going down this wormhole . . . it was like, *Oh my G-d, I think this is it.*[1] I think that food could change a lot for me. And that's where it all started."

It's never too late to do a 180, but flexibility is time's gift to the young. Alex seized her epiphany and set her sights overseas, deciding to quit a dead-end boyfriend and job and take an extended trip with a friend. She had heard about Worldwide Opportunities on Organic Farms, or WWOOF (like the dog bark), as it's called—a service exchange program where people can volunteer on small, sustainable farms in more than 130 countries. Since Alex's foray, the organization has faced criticism for labor issues: Volunteers are supposed to work four to six hours a day in return for room and board, but tales abound of overworking and underfeeding. (Like anything else in life, just do your research.) In 2008, however, none of this had come

---

1 Hopefully this book will never be thrown out, but just in case you're wondering: It is Jewish tradition to never write out the name G-d, out of respect, in case what it's written on is discarded. The Torah prohibits the erasure of G-d's name.

to light. Traveling in Europe, Alex split off from her friend and set off to find a WWOOF farm—landing with a family of cheesemakers sixty miles north of Rome in the untraveled agrarian village of Montefiascone, Italy.

"I stepped onto this farm. It was in Italy, so it was magical by default," she laughed. "I had my little backpack and I couldn't speak Italian." She wound up staying with her host family—a Sicilian man, his German wife, and their son—for four months. "They had forty sheep. . . . They milked their sheep to make cheese. And in exchange for cheese, they would trade with the olive oil folks and graze their sheep flock under [their] olive trees. It was like nothing I had ever experienced." She and the family became close, and they asked Alex how she felt about a solo farm stay for a month while the family headed off on vacation. (Glorious sounding, but note: not the guideline WWOOF arrangement.) Alex remembered her response of pure bliss: "I was like, *Yeah! That sounds great!*"

She couldn't drive anywhere. She had a slider cell phone with no phone plan. "I had these sheep, and the family left me a bin of flour and a wheel of cheese. I had to forage for vegetables and fruit, and I loved it!"

When Alex returned to Arizona, she contemplated her options and decided, "I'm not doing anything else [but farming] ever again. I'm the healthiest I've ever felt. I feel joy for the first time in so long." My eyes were welling up on the other end of the line as she said this. By this point in Alex's story, she was weaning off the medication and had long forgone illicit drugs. "It was a really quick flip from being on the standard American diet and being so inflamed and in so much pain, plus depressed and [anxious]."

A healing approach to food, she witnessed through Italy's ancestral food culture, went hand in hand with a holistic way of life. "I just never went back to anything else. That was it."

KNOWING HOW WAY LEADS on to way, as poet Robert Frost wrote, Alex's story accelerates from here. She worked at a plant nursery for a day until she witnessed the copious chemical use, then landed a job working Phoenix-area farmers' markets for a hydroponic tomato seller. That, too, proved compromised (months in, she spied "farm-fresh" produce coming off the truck with PLU—price look-up—code stickers on it) and resolved to enroll in a sustainable farmer training program, ultimately spending a year at the Organic Farm School on Washington State's Whidbey Island. Because Alex had gone to rehab instead of college, her parents helped with the $8,500 tuition. "They were really, really generous," she said. But the money came with a stipulation: "They said, 'We will help you go, but this is the only time we're going to help you pay for any educational experience.' And it was invaluable. It was incredible."

Afterward, Alex came back to Arizona and traveled around, trying out farm life. She rounded out her skills and earned income at Rattlebox Farm, an organic vegetable farm, as well as Rezona-tion Farm, an egg and honey farm. She has the beekeeping suit pics to prove it. But her most transformative experience was a work exchange at Bean Tree Farm in the Tucson foothills, designing earthworks[2] and water-harvesting systems around food-bearing native desert plants. The "farm," in fact a United Plant Savers Botanical Sanctuary, is part of a community of five rammed-earth solar homes on twenty acres of Sonoran Desert.

"[It's] called the food forest because, believe it or not, almost everything in the Sonoran Desert is either edible or medicinal

---

2 Terraces, swales, dams, ponds, and *hügelkultur* (an ancestral Eastern European way of building mound-like raised beds from decaying wood and plant debris; the word means "hill culture") are all examples of earthworks employed in sustainable agriculture.

for humans," Alex said. Saguaro and prickly pear fruit, mesquite pods, cholla buds—these are a common few of more than 500 species used immemorially by the Tohono O'odham people.[3] "Their food system is ancient; the Native folks of that area are still there farming [and gathering] in the same way. It was amazing to witness." This experience would prove to be the incubator for reclaiming an age-old food culture of her own, right back where she had started.

⁂

WHEN I FIRST TOLD my mom about the Jewish Farmer Network, an organization connecting more than 2,000 Jewish farmers and growers around the world, she delivered a response I was about to learn is ubiquitous, most especially among Jews: *Jewish? Farmer? Isn't that an oxymoron?*

"That's something we hear again and again from [beginning Jewish] farmers in our network," said Shani Mink, Jewish Farmer Network's executive director and cofounder, as well as a seasoned farmer herself. "They go home to their families and say, 'I'm getting a job on a farm; I'm really excited!' Their families say to them: 'Jews don't do that. Jews don't farm.'"

It was the last day of Hanukkah in December 2020, and I had called Shani from our then home in Oregon, where I was about to apply to a farmer training program and thus become an ostensible oxymoron. Shani had spent the pandemic's outset working on the farm of Jewish Farmer Network's other cofounder, SJ Seldin, out-

---

3 The O'odham people (Tohono is the desert designator), descendants of the bygone Hohokam people, today encompass four federally recognized tribes: the Tohono O'odham Nation, the Gila River Indian Community, the Ak-Chin Indian Community, and the Salt River Pima-Maricopa Indian Community.

side Asheville, North Carolina, followed by another farm in New York on Long Island's East End.

In Shani's circles, "Jewish farmer" isn't a contradiction. She was one of the first food and farming fellows with Adamah (formerly Hazon), a nonprofit focused on environmental education in Jewish communities. And in Israel, the collective farming community known as a kibbutz is ubiquitous. My own father spent a summer on one as a teen in the 1960s. But still, in my mind—and the psyches of most American Jews and non-Jews alike—Jews are doctors, businesspeople, and Hollywood comedy writers, which is why I was blown away when I first heard what Shani was creating and wanted to speak with her. She started off by unpacking the "Jews don't farm" myth. "If you think about the [now tainted] Woody Allen Jewish trope—"

"—that Jews are nerdy, bookish," I interrupted. (Linguist Deborah Tannen calls this typical Jewish conversation style "cooperative overlapping"; I usually try to keep my interposing in check when interviewing.)

"—nebbish, if you will," completed Shani, to my laugh. "It's not true because our tradition is a fully agrarian tradition, and people just don't know that because we're caught up in this cloud of stories."

These stories, though, stem from the real history of the Jewish people: a largely diasporic population prevented from owning land and property, as well as entering professions open to others. In czarist Russia, for instance, Ashkenazi Jews such as my ancestors could neither buy land nor live in agricultural communities. They were restricted to the Pale of Settlement (essentially a giant Jewish ghetto) or ousted to shtetls on impoverished land. The "Jews run Hollywood" echoism arose from the same discriminatory history: Antisemitism blocked Jewish people from entering many established industries in the early twentieth century, but they found an entrée in the lowly field of vaudeville, which evolved into the just-as-spurned burgeoning film industry. My last name, Grayson, in fact, was Goldstein: My grandfather changed his name to bypass bigotry on the trading floor of the New York Stock Exchange.

Even today, "A lot of us—especially Ashkenazi Jews—are walking around with Nazi-era antisemitic propaganda in our brains: the idea that Jews are weak," Shani said.

The story Jewish Farmer Network seeks to strengthen is that Judaism—not just a religion, but a culture and a peoplehood—is an age-old agrarian tradition. "The entire Jewish calendar is built around the cycles of the harvest," she explained. Jewish holidays, in fact, waypost the timetable: Passover recounts the Exodus from Jewish slavery in Egypt, but it also signals the start of the grain harvest. Shavuot commemorates when G-d handed down the Torah to Moses at Mount Sinai and marks the end of the grain reaping and beginning of the fruit gathering.

Then there are the agricultural ethics interwoven throughout the Torah and ancient dictated oral law of the Talmud—relevant laws on sustainability and food justice that are, in fact, millennia old: *Pe'ah* ("corner") dictates edges of the field be left for those in need. *Leqet* ("a gleaning") mandates leaving missed harvest for the poor. *Shmita* ("release") commands letting the land lie fallow every seven years—a practice more necessary than ever on an earth now stripped of a third of its topsoil.

"And if we're not going to till our land and grow during the seventh year, how are we going to survive?" posited Shani. "Oh, right: We have to invest in perennials and think long-term. It all begins to connect into this beautiful picture of how to be in relationship with the land in a way that is regenerative for us, regenerative for the soil, and also regenerative for the community at large, both ecological and social communities."

This was a renewable agriculture even in antiquity. The Roman colonies of Cuicul, Timgad, Sousse, and Jerash; the Nabatean city of Petra; the Syrian settlement of Antioch; the Phoenicians in Lebanon—all of these settlements collapsed after their peoples exhausted their lands, detailed geomorphologist David R. Montgomery in his book *Dirt: The Erosion of Civilizations.* The exception was the ancient Israelites, who practiced shmita, rotated their crops, harvested rainwater, and made straw-manure compost. Their stone terraces "still held soil after several thousand years of cultivation," he wrote.

Like the earliest Jewish farmers, a sustainable existence was Alex Rosenberg-Rigutto's goal. Blame it on inherited memory of wandering the desert and subsisting on manna; though she was awed by Arizona's native food heritage, there was another takeaway from her time farming there: "There's no water here!" Drought, extreme heat, and a swelling population are jeopardizing what scarce water remains in the state. So Alex returned to Michigan (though a good 300 miles from Detroit) to the town of Pellston at the top of the Lower Peninsula. She found a new gig at Open Sky Farm, a four-acre organic vegetable and flower farm started by Sam and Susan Sharp as their retirement project. ("We decided a while back that small-scale farming is one of the best ways to affect [*sic*] social and environmental change," their website states. I have to say I agree.)

"I spent a lot of time up there in northern Michigan, really becoming acquainted with *Who else is here? Who are the trees? Who are the plants? Who are the insects? What is this land? What is its history?* I wanted to come back to Michigan after [my] traumatic experience [there] in a very gentle and intentional way," Alex said.

A year later, friends from her Whidbey Island school days sent her a job posting: Michigan's Tamarack Camps was looking for a founding director for a new farm on its 1,100 forested acres in Ortonville. The camp was founded 120 years ago on Detroit's Belle Isle Park as a nature retreat for inner-city Jewish immigrant kids. Her friends had coincidentally met at Tamarack and gotten married there, and they knew Alex was back in Michigan. "Help them build it!" they begged Alex.

Alex was hesitant to apply at first—in part because she wasn't sure she was qualified to lead such a significant project. But there was something else holding her back: She was scared to rejoin the "disconnected" Jewish community of her childhood memories. "It was so painful for me as a youth. I had no [better] example. I didn't know what the Jewish Farmer Network was; I had no idea that Jewish agrarianism was a field in itself," said Alex, who was later introduced to the network.

Alex was echoing the same misconceptions I had spoken about with Shani. I partly blame our dispersed food culture. "I have friends who are Chinese American farmers and they're growing traditional Chinese vegetable varieties to be sold back to their communities for more authentic cooking. The same is not really true for Jews," Shani said.

"I wish our food was as good as traditional Chinese food," I added. (I give you: kugel?) "You've just gotta hang out with more Mizrahim and Sephardi," Shani said. "Then you won't complain."

Shani, in her youth, hadn't known about Jewish agricultural tradition either. She was raised in a working-class modern Orthodox family in the affluent Jewish suburb of Livingston, New Jersey—a community not unlike the Detroit suburbs Alex grew up around. But as a kid, Shani sensed something was amiss: "My family would sit in this sukkah [the temporary shelter Jews dwell in during the holiday of Sukkot], with plastic grapes and plastic fruit hanging . . . and I'd be like, *This is a fall harvest festival. Why are we putting plastic fruit up in our sukkah? Why can't we actually connect with what's being harvested right now?*

Shani followed that interest in local food to her first farm job at the age of eighteen, scheduling all her classes in the afternoon so she could farm through college. But "Jewish" and "farmer" were disparate until she went to work at Eden Village Camp, a Jewish organic farm camp in Putnam Valley, New York. During her staff training, a camp leader casually mentioned, "Judaism is one of the oldest earth-based traditions still being practiced today," Shani recalled. "And I was like, 'What?' . . . While we were talking, this whole world was opening up to me where I [realized]: Oh, wow. All of this earth connection that I had been looking for in *other* places is all here, rooted in my tradition and the traditions of my people, my ancestry."

Clearly this concept struck a chord, just as cultural reclamation through farming has for other dispossessed communities and anyone longing to recover their roots: In 2016, Shani found herself chatting in a group of thirteen farmers at a Jewish food and farming conference about the loneliness of often being the sole Jew in agricultural settings. The next day, she and her Jewish Farmer Network cofounder, SJ Seldin, started a Facebook group to connect Jewish farmers; two days later, 200 people had joined.

Today, there are more than 2,000 Jewish Farmer Network members—composed of mostly current but also aspiring regenerative farmers—and Shani has helped develop the resources to connect and support them, including virtual classes in Jewish sustainable agrarian practices and principles of food justice, digital holiday guides, a yearly conference, and a newsletter with regularly posted farming jobs. In the future, she hopes to tackle the obstacle of land access—arguably the biggest issue for new small farmers of all backgrounds—through a land-matching and stewardship program. It would pair Jewish farmers looking for land with Jewish institutions that own it: synagogues, camps, day schools, and more. This is a model with cross-cultural power: The nonprofit Agrarian Trust is working to connect sustainable farmers with millions of acres of church-owned land through its FaithLands project.

As for Alex, who is now a friend of Shani's, I think you know the ending: She set aside her fear, took a leap of faith, and decided to apply for the Farber Farm job. "Ten interviews and four months later, I was hired!" she said.

Today, on what was swampy abandoned horse pasture, she and her crew grow more than a hundred varieties of vegetables, fruits, herbs, flowers, cover crops, and of course, native perennial plants using organic growing practices. There are chickens, ducks, rabbits, and honeybees. There's a learning center and an outdoor pavilion; there are in-ground gardens, raised-bed accessibility gardens, a greenhouse, a hoop house, even a cob oven—all with the purpose of growing produce for the community and reconnecting new generations of Jewish kids to a venerable food and farming history.

"Here we were, the first shepherds ever.[4] But [most of us] can't recall that because . . . we're not shown the examples," Alex said. Because of the Jewish Farmer Network, "I was able to unpack what I learned as a kid and learn about Judaism in a completely different way that felt nourishing and so right and something that I had been hungry for forever."

Alex and I were chatting in September 2022, on the second day of Rosh Hashanah, the Jewish New Year that launches the High Holy Days, a time of renewal and contemplation. *And* it marks the upcoming end of the fall harvest, a time to thank G-d or pray for help through "the white blanket" of a Michigan winter. Back in Los Angeles, I had just harvested a Japanese cucumber from my garden to slice on my cultural mash-up breakfast sandwich of farmer cheese on whole kernel rye. I wished that back in my twenties I had aligned my lifestyle with my values as clearly as Alex and Shani, but I felt thankful for what I had reclaimed on this journey: a dilapidated side yard I've dubbed Coyote Hill Farms, now packed with grow

---

4 Clarification: *among* the first shepherds. Recently analyzed sheep and goat remains of a Neolithic site in modern-day Kyrgyzstan date herding in central Asia to 8,000 years ago.

bags of vegetables and fruit trees, and a robust Jewish community in Los Angeles in which to raise my kids—the now meaningfulness of which I never could have imagined.

Alex had a lot to reflect on too: She had a milestone birthday that December—"I'm almost thirty. Oh my G-d, it's huge," she said—and was about to celebrate her first wedding anniversary: She had gotten married the year before to fellow farmer and foodie Rick Rigutto, who was now helming the on-site farm at hyperseasonal dining destination Sylvan Table in Sylvan Lake, Michigan. The two were also embarking on a farmstead of their own: Northwoods Farmstead and Skill Center on 10.5 acres in backwoods Hersey, Michigan, where they plan to grow mixed provisions, host immersive skills classes, and eventually offer farm stays.

Alex's story seemed to be a miracle. She had found her way to such an authentic way of life, and I asked her how it felt. There were challenges, of course: "If you have an implement stuck on your tractor, and it's eight o'clock at night, and you're tired, and the sun is going down, and this thing won't come off—you're gonna get your chain, hook it up, get your car, and pull the thing off, you know what I mean? I'm not able to just call somebody to help me like, *There's an IT guy for that*," she laughed.

"But oh my gosh, in the day-to-day—this is going to sound so tacky—I just feel very blessed every single day that I get to wake up, and live my values, and eat food that I've grown—and that I get to wake up next to somebody that I love so much, who is also so aligned. It's just this overwhelming feeling of gratitude."

ACKNOWLEDGMENTS

SOME BOOKS YOU DREAM UP and can't wait to write, and others grab hold of you and won't let go until you deliver what they have to say. This book was my own hero's journey, and I am so grateful to everyone who made it possible and supported me along the way:

My first thank-you is to my agent, Mel Parker, who has been my champion and sounding board for nearly a decade now. His wisdom and elegance navigating any situation helped me see the way forward when I myself could not.

To Carys Wilkins, Benji Nagel, and their family, who opened their home, hearts, and livelihood and entrusted me to capture the remarkableness of it all, thank you. My biggest motivation and hope was that I might do you justice. I was driven by that same call for all the extraordinary farmers, graziers, and food activists featured and named in this book, and I send them my deepest appreciation and admiration. To Cedar, Em Blood, and Elliott Blackwell, who didn't know when they signed up for a farming internship that their crew later would include a trailing journalist, thank you for your awesomeness and for making those months one of the most fun experiences of my life. This project also would not have existed without the groundbreaking work of Rogue Farm Corps and the support of Gia Matzinger, Matt Gordon, and Abigail Singer; I am indebted. I additionally hope that Niki Timm-Branch and Becky McColl at Central

Oregon Locavore are as delighted as I am that they unknowingly steered me toward this journey. Katie, the conduit for my lightning bolt, I send my gratitude to you wherever your passion and travels have taken you.

Several folks shared ideas with me, connected me with sources, or graciously took the time to chat with me before I knew where any of this was going: Peter Michael Bauer, Mellissa Berry, Brittany Cole Bush, Rufus Haucke and Joy Miller, Lisa Valinsky, and Hazel Ward—I have not forgotten you, and I thank you. Natalie Bogwalker, Ariel Greenwood, and Shani Mink were among these early conversations, and I deeply appreciate their willingness not only to share their stories and knowledge but also to stick with me over the years of this book. I also want to acknowledge all the additional farmers and experts who added invaluable insight but whose names are either not mentioned or mentioned only briefly in the text, including Jordan Amaker, Daron Babcock, Lindsey Barrow and Olivia Myers, Imani Black, Suzie Flores, Suzanna Goldblatt Clark, Ben Larson, Chloe Lieberman, Susan Lightfoot Schempf, Anthony Natoli and Ellen Waldrop, Jeff Privette, Japhety Ngabireyimana, Jeff Snyder, Jeff Siewicki, Brian Wheat, and JR Zumwalt.

Tai Power Seeff also offered early inspiration; she deserves her own line here as well as my enduring thanks. I have never known a kinder or more generous artist, and her breathtakingly beautiful photos of Mahonia Gardens grace the jacket and interior of this book.

I am beyond grateful to my wonderful editor, Isabel McCarthy, and the entire team at Countryman Press for being as called to this project as I was from the get-go. Isabel left too soon (to take another position), but her judicious edits brilliantly clarified and strengthened the work. Thankfully, I have been in the best possible hands with ace editor, James Jayo, who expertly steered me toward the finish line. I also deeply appreciate the skill, support, and behind-the-scenes efforts of Ann Treistman, Jess Murphy, Maya Goldfarb,

Julia Druskin, Zach Polendo, and Devorah Backman. Thank you to Sarah Bibel, for a cover that is absolute perfection. And Kathryn Flynn deserves a monumental shout-out for a flawless copyedit—I think only a former proofreader could fully appreciate the profundity of her catches.

This book also would not have been possible without the listening and support of my dear friends, Meghann Haldeman, Rachel Kash, and Rachel (Sat Siri) Dougherty. Sarah and Noah Grayson infuriatingly told me I could write the book first and then go to med school. Katherine Grayson, my mom, is my most-devoted reader. Lindsay and Daniel Libman are my foodie soulmates and an endless source of support and ideas. Ron Roman kept my glass full in Charleston when the going got tough. And Sam Rubin, Susan Handelman, and Joie Scott-Poster made it a party when the manuscript was in.

To my beautiful husband, Matthew Libman, and daughters, Isobel and Mika—together, my everything—there are no words. All I can say is that I'm excited for the next chapter.

RESOURCES

Immersive Training Programs in Regenerative Agriculture for Aspiring Farmers

Adamah Farm Fellowship (adamah.org/for-teens-youth-adults/adamah-farming-fellowship)

Center for Land-Based Learning California Farm Academy (landbasedlearning.org/farm-academy-beginning-farmer.php)

FarmaSis (www.thefarmasis.com)

Hawthorne Valley Farm One-Year Diversified Farming Apprenticeship (farm.hawthornevalley.org/apprenticeship-overview)

Maine Organic Farmers and Gardeners Association Apprenticeship Program (www.mofga.org/apprenticeship)

Occidental Arts & Ecology Center Garden Internship (oaec.org/garden-internship)

The Organic Farm School on Whidbey Island (organicfarmschool.org)

Quivira Coalition New Agrarian Program (quiviracoalition.org/newagrarian-dnu)

Rodale Institute Farmer Training Program and Veteran Farmer Training Program for military veterans (rodaleinstitute.org/education/training-programs)

Rogue Farm Corps (www.roguefarmcorps.org)

The Sustainable Agriculture Education Association created this list of colleges and universities that offer programs in sustainable agriculture: sustainableaged.org/academicdegreeprograms.

Additional Organizations

Agrarian Trust (www.agrariantrust.org)

American Farmland Trust (farmland.org)

Asian American Farmers Alliance (farmermai.com)

Black Food Sovereignty Coalition (blackfoodnw.org)

Central Oregon Locavore (centraloregonlocavore.org)

Cohousing Association of America (www.cohousing.org)

Community Supported Grocery (www.communitysupported
grocery.com)

Dream of Wild Health (dreamofwildhealth.org)

Ecological Farming Association (www.eco-farm.org)

Farmer's Footprint (farmersfootprint.us)

Fibershed (fibershed.org)

Food Tank (foodtank.com)

GMOScience (gmoscience.org)

Greenhorns (greenhorns.org)

GreenWave (www.greenwave.org)

Holistic Management International (holisticmanagement.org)

Jewish Farmer Network (www.jewishfarmernetwork.org)

Kiss the Ground (kisstheground.com)

Minorities in Aquaculture (www.mianpo.org)

National Gardening Association (garden.org)

National Young Farmers Coalition (www.youngfarmers.org)

Native American Food Sovereignty Alliance (nativefoodalliance.org)

North American Traditional Indigenous Food Systems (natifs.org)

Queer Farmer Network (www.queerfarmernetwork.org)

Regeneration International (regenerationinternational.org)

Regenerative Organic Alliance (regenorganic.org)

Seed Savers Exchange (seedsavers.org)

Slow Food International (www.slowfood.com)

Soil Health Institute (soilhealthinstitute.org)

Soul Fire Farm (www.soulfirefarm.org)

Southeastern African American Farmers' Organic Network
     (saafon.org)
Worldwide Opportunities on Organic Farms (wwoof.net)

For further reading on regenerative farming and ways to get involved, this compilation by Courtney White at Project Regeneration is the best I've seen: regeneration.org/nexus/regenerative-agriculture.

NOTES

Introduction

9   **extractive agriculture primed a fire:** Katie Rodriguez, "How Centuries of Extractive Agriculture Helped Set the Stage for the Maui Fires," *Civil Eats*, August 23, 2023.

12   **400 million acres:** The Oakland Institute, *Down on the Farm: Wall Street: America's New Farmer*, 2014.

12   **nearly half of all farmland:** US Department of Agriculture, *Farms and Land in Farms 2022 Summary*, USDA National Agricultural Statistics Service, February 2023.

12   **around a third of the world's food:** Sarah K. Lowder, Marco V. Sánchez, and Raffaele Bertini, "Which farms feed the world and has farmland become more concentrated?" *World Development* 142 (June 2021), 105455.

12   **up to 90 percent of Americans:** Andrew Zumkehr and J. Elliott Campbell, "The potential for local croplands to meet US food demand," *Frontiers in Ecology and the Environment* 13, no. 5 (June 2015): 244–48.

13   **Project Regeneration:** Courtney White, "Regenerative Agriculture," Project Regeneration, regeneration.org.

13   **for the past ten years:** US Department of Agriculture, *2022 Census of Agriculture*, USDA National Agricultural Statistics Service, February 2024; US Department of Agriculture, *2017 Census of Agriculture*, USDA National Agricultural Statistics Service, Census of Agriculture Historical Archive, April 2019.

13   **the Great Resignation:** Joseph Fuller and William Kerr, "The Great

Resignation Didn't Start with the Pandemic," *Harvard Business Review*, March 23, 2022.

13 **Two hundred years ago ... even a hundred years ago:** "Farming in the US (Timeline)," *American Experience*, PBS, PBS.org.

13 **that number stands at 1 percent:** USDA, *2022 Census of Agriculture*.

Chapter One. Mahonia Gardens

18 **fifty-eight years old:** US Department of Agriculture, *2022 Census of Agriculture*, USDA National Agricultural Statistics Service, February 2024.

19 **peasant farmers' vegetable plots:** "The history of market gardening," *Alimentarium*, alimentarium.org (accessed January 26, 2022).

19 **Māori gardeners:** Maggy Wassilieff, "Market gardens and production nurseries: History of market gardening," *Te Ara: The Encyclopedia of New Zealand*, November 24, 2008, teara.govt.nz (accessed January 26, 2022).

20 **doubled between 2000 and 2020:** USDA Economic Research Service, "Average US farm real estate value, nominal and real (inflation adjusted), 1970-2020," www.ers.usda.gov/webdocs/charts/55910/farmrealestatevalue 2020_d.html?v=6230.

20 **the average cost of an acre:** US Department of Agriculture, *Land Values 2022 Summary*, USDA National Agricultural Statistics Service, August 2022.

20 **463 acres:** USDA, *2022 Census of Agriculture*.

20 **$1.5 trillion:** Congressional Research Service, *Farm Bill Primer: Budget Dynamics*, CRS Reports, December 4, 2023.

23 **Oregon's high desert:** Jeff LaLande, "High Desert," *Oregon Encyclopedia* (The Oregon Historical Society), July 12, 2023.

23 **steppe:** Daniel Costa, "steppe: grassland," *Encyclopedia Britannica*, britannica.com (accessed Oct. 19, 2021).

23 **162 days of sunshine:** "Climate," Redmond Economic Development Inc., rediinfo.com/climate.

25 **ammonium phosphate:** US Food and Drug Administration, "21CFR184.1141a," Department of Health and Human Services, November 18, 1983 (updated December 22, 2023).

28 **over eleven moves:** US Census Bureau, "Calculating Migration Expectancy Using ACS Data," Census.gov, December 3, 2021.

31   **$2 million an acre:** Richard Florida, "The Staggering Value of Urban Land," *Bloomberg*, November 2, 2017.

32   **food that your great-great-grandmother would recognize:** Michael Pollan, "Unhappy Meals," *New York Times Magazine*, January 28, 2007.

32   **food hubs:** USDA Agricultural Marketing Service, *Regional Food Hub Resource Guide*, US Department of Agriculture, April 2012.

33   **Grandview Fire . . . Bootleg Fire:** Northwest Interagency Coordination Center, *Northwest Annual Fire Report 2021*, March 8, 2022, gacc.nifc.gov/nwcc.

Chapter Two. Wild Abundance

38   **permaculture:** Bill Mollison and David Holmgren, *Permaculture One: A Perennial Agriculture System for Human Settlements* (Melbourne: Transworld Publishers, 1978).

42   **daylilies:** Hank Shaw, "Dining on Daylilies," *Hunt Gather Cook*, June 29, 2010 (updated May 13, 2020); Olallie Daylily Garden, "Cooking and eating Daylilies (Hemerocallis)," daylilygarden.com.

42   **the Can Masdeu squat:** "Nuestra Historia," Vall de Can Masdeu, canmasdeu.net.

43   **for nearly a billion people:** Zareen Pervez Bharucha and Jules Pretty, "For millions, wild food is no fad but a matter of life or death," *The Conversation*, March 31, 2014.

43   **ethnobotanist Frank Cook:** Robin Harford, "In memory of Frank Cook (1963–2009)," *Eatweeds*, eatweeds.co.uk; Arnyce Pock, "Medical Report of Death," *Plants and Healers*, posted September 9, 2010.

46   **dagga:** Baudry N. Nsuala, Gill Enslin, and Alvaro Viljoen, "'Wild cannabis': A review of the traditional use and phytochemistry of *Leonotis leonurus*," *Journal of Ethnopharmacology* 174 (November 4, 2015), 520–539.

46   **legal and uncontrolled in the US:** US Drug Enforcement Administration, "Controlled Substances: Alphabetical Order," DEA Diversion Control Division, December 14, 2023; Louisiana State Legislature, "Revised Statute 40:989.2 – Unlawful production, manufacturing, distribution, or possession of prohibited plant products; exceptions," 2018.

51   **eggs are scarce due to an avian flu outbreak:** USDA Animal and Plant Health Inspection Service, "2022-2024 Confirmations of Highly Patho-

genic Avian Influenza in Commercial and Backyard Flocks," US Department of Agriculture, January 2, 2024.

51 **five of every six farms in the world . . . 35 percent of the world's food:** FAO Newsroom, "Small family farmers produce a third of the world's food," Food and Agriculture Organization of the United Nations, April 23, 2021.

52 **"gangsta gardener" Ron Finley:** "Who We Are: The Ron Finley Project," Ron Finley Project, 2024.

52 **"Right to garden laws" have recently been passed:** Katherine Kornei, "Only Two States Have Passed 'Right to Garden' Laws. Will Others Follow?" *Civil Eats*, August 20, 2022.

53 **The Victory Garden movement:** Tejal Rao, "Food Supply Anxiety Brings Back Victory Gardens," *New York Times*, March 25, 2020; "Victory Gardens," *New York Times*, May 6, 1944.

54 **what really led to the rationing and food shortages:** J. Walton, "Reclaiming Victory Gardens from Our Racist History," *Green America*, April 21, 2020.

54 **internment was . . . an agricultural land grab:** A.V. Krebs, "Bitter Harvest" (Opinion), *Washington Post*, February 2, 1992.

Chapter Three. Ambler Farm

59 **whose alumni included:** Jonathan Saltzman and Jenn Abelson, "Former students call for inquiry into assault claims," *Boston Globe*, May 9, 2016; "Notable Alums," The Fessenden School, fessenden.org.

59 **ready young men for Yale:** "Brief History," The Hotchkiss School, hotchkiss.org.

63 *aigamo:* "Farming Rice with Ducks," Web Japan, Ministry of Foreign Affairs of Japan, October 22, 2002.

63 **"farmer to the stars" Annie Farrell:** Amy Kalafa, "Sustainable Connecticut: Farmer to the Stars," *Cottages & Gardens*, April 7, 2011.

63 **$5.9 million:** Pat Tomlinson, "Wilton's Millstone Farm finds new owners," *The Hour*, December 9, 2016.

63 **"a lavish, country-style ceremony":** "What's Happening: October 2022," Eurodressage, October 19, 2022.

66 **95 percent of farms:** US Department of Agriculture, *2022 Census of Agriculture*, USDA National Agricultural Statistics Service, February 2024.

66   **"family farms":** Christine Whitt, US Department of Agriculture, "A Look at America's Family Farms," USDA Economic Research Service, Resource and Rural Economics Division, January 23, 2020.

66   **100 biggest landowners:** Dave Merrill, Devon Pendleton, Sophie Alexander, Jeremy C.F. Lin, and Andre Tartar, "Here's Who Owns the Most Land in America," *Bloomberg*, September 6, 2019.

66   **Bill Gates is now the largest owner of private farmland:** Ariel Shapiro, "America's Biggest Owner of Farmland Is Now Bill Gates," *Forbes*, January 14, 2021.

67   **such as shoemaking and distilling:** "Wilton Town History," The Town of Wilton, Connecticut, wiltonct.org.

68   **Native peoples of the region:** "Map of the Week! Connecticut Tribes Circa 1625," *Outside the Neatline*, University of Connecticut, August 10, 2009.

68   **selectively harvested the landscape:** Tobias Glaza and Paul Grant-Costa, "Breaking the Myth of the Unmanaged Landscape," Connecticut History.org, February 6, 2022.

71   **Garment work:** Howard M. Sachar, *A History of the Jews in America* (New York: Vintage, 1993).

72   **40 percent:** US Department of Agriculture, *2014 Tenure, Ownership, and Transition of Agricultural Land (TOTAL) Survey*, USDA Economic Research Service (updated February 12, 2024).

73   **$15,000 in farm sales:** Connecticut Department of Agriculture, "Agricultural Property Tax Exemptions and Abatements," ct.gov.

74   **64 percent of small farmers:** Brian Barth, "By the Numbers: The State of Today's Independent Farmer," *Modern Farmer*, June 29, 2018.

74   **largely undocumented:** Miriam Jordan, "Farmworkers, Mostly Undocumented, Become 'Essential' During Pandemic," *New York Times*, April 2, 2020.

74   **"digital agriculture":** Marco Bellizi, "Vandana Shiva on how we risk becoming 'obsolete technology'," *Vatican News*, November 20, 2020.

Chapter Four. Round Table Farm

81   **Hardwick's genealogical registry:** Lucius R. (Lucius Robinson) Paige, 1802–1896, *History of Hardwick, Massachusetts: with a genealogical register, 1883*, Rare Book Collections, Special Collections and University Archives, University of Massachusetts Amherst Libraries.

82   **More than a third:** US Department of Agriculture, *2022 Census of Agriculture*, USDA National Agricultural Statistics Service, February 2024.

82   **Only a quarter:** US Department of Agriculture, "Farmland Ownership and Tenure," USDA Economic Research Service, May 16, 2022.

82   **Senator Josh Hawley:** "S.684—118th Congress (2023–2024): This Land Is Our Land Act," Congress.gov, March 7, 2023.

82   **Chinese ownership of American farmland:** Emily Washburn, "How Much US Farmland Does China Really Own? More than Bill Gates—and Less than 17 Other Countries," *Forbes*, March 1, 2023.

87   **two-thirds of the state's dairy farms:** Lisa Luciani, "Dairy Farms in Massachusetts, and Across New England, Are Evaporating," *Land for Good*, August 16, 2021.

87   **LGBTQ adults:** American Psychiatric Association, "Diversity & Health Equity Education: Lesbian, Gay, Bisexual, Transgender, and Queer/Questioning," Psychiatry.org.

87   **and farmers, who both suffer:** Rosalie Eisenreich and Carolyn Pollari, "Addressing Higher Risk of Suicide Among Farmers in Rural America," National Rural Health Association Policy Brief, NRHA.

88   **US government subsidies:** Stephanie Luiz, "My Beef with Dairy: How the US Government Is Bailing out a Dying Industry," *Northeastern University Political Review*, May 16, 2020.

88   **led 10,000 dairy farms to shutter:** Lisa Luciani, "Dairy Farms in Massachusetts."

93   **$2.6 million:** Jim Russell, "Cannabis growing facility planned at Hardwick farm," *MassLive*, March 28, 2022.

93   **a thoroughbred racetrack and gambling facility:** Jon Chesto, "In Hardwick, a vote on horse racing divides the town," *Boston Globe*, January 6, 2023.

93   **800 farms and 68,000 acres:** "Agricultural Preservation Restriction Program (APR)," Beginning Farmer Network of Massachusetts.

94   **rejected the plan in a referendum:** Kevin Koczwara, "Hardwick votes down racetrack," *Worcester Business Journal*, January 9, 2023.

94   **87 percent white:** US Census Bureau (2022), *American Community Survey 5-year estimates,* retrieved from Census Reporter profile page for Hardwick town, Worcester County, MA.

100 **people just have this land:** Yuta Kumamoto, "Projects to lease abandoned farmland in west Japan urban areas proving popular," *The Mainichi*, January 19, 2020.

100 **focuses its crop subsidies:** Tara O'Neill Hayes and Katerina Kerska, "Primer: Agricultural Subsidies and Their Influence on the Composition of US Food Supply and Consumption," American Action Forum, November 3, 2021.

100 **Japan directly subsidizes—gasp!—vegetables:** Kenzo Ito and John Dyck, "Vegetable Policies in Japan," *Electronic Outlook Report from the Economic Research Service,* USDA Economic Research Service, November 2002.

100 **4.5 percent obesity:** "Obesity Rates by Country, 2024," World Population Review, 2024.

100 **42 percent:** Craig M. Hales, Margaret D. Carroll, Cheryl D. Fryar, et al, "Prevalence of Obesity and Severe Obesity Among Adults: United States, 2017–2018," *NCHS Data Brief*, no. 360, February 2020.

101 **grow produce as a side job:** Akiba, "The Food Chain: Japanese Agriculture—From Farm to Supermarket," *Hackerfarm*, April 21, 2019.

101 **Nipmuc:** Amy Tikkanen, "Nipmuc," *Encyclopedia Britannica*, britannica .com (accessed May 10, 2023).

105 **other young Asian Americans:** Allison Park, "Young Asian Americans turn to farming as a means of cultural reclamation," *NBC News*, October 28, 2019.

106 **early horticultural societies:** Rolf Derpsch, "History of Crop Production, With & Without Tillage," *Leading Edge, The Journal of No-Till Agriculture* 3, no. 1 (March 2004), 150–154.

106 **4000 BC:** Amy Bogaard, Mattia Fochesato, and Samuel Bowles, "The farming-inequality nexus: new insights from ancient Western Eurasia," *Antiquity* 93, no. 371 (October 2019): 1129–1143.

106 **"self-scouring" steel plow:** "4 Inventors Who Advanced American Agriculture," National Inventors Hall of Fame (updated February 29, 2024).

106 **John Froelich:** "Biography of John Froelich from Iowa Inventors Hall of Fame Pamphlet, 1994," State Historical Society of Iowa, 1994 (accessed May 10, 2023).

106 **degraded a third of the earth's topsoil:** Intergovernmental Technical Panel on Soils, *Status of the World's Soil Resources, Main Report*, Food and Agriculture Organization of the United Nations, 2015.

106 **catastrophic levels of carbon:** International Union for Conservation of Nature, "Land degradation and climate change," IUCN Issues Brief, November 2015.

106 **95 percent of our food:** Susan Cosier, "The world needs topsoil to grow 95 percent of its food—but it's rapidly disappearing," *Guardian,* May 30, 2019.

106 **more than 90 percent:** Global Soil Partnership, "Saving our soils by all earthly ways possible," Food and Agriculture Organization of the United Nations, July 27, 2022.

106 **soil degradation . . . soil erosion:** FAO Soils Portal, "Soil degradation," Food and Agriculture Organization of the United Nations (accessed December 13, 2023).

107 **16 percent:** Hannah Ritchie, "Do we only have 60 harvests left?" *Our World in Data,* January 14, 2021.

107 **on the soil microbiome:** Judith Kraut-Cohen, Avihai Zolti, Liora Shaltiel-Harpaz, et al, "Effects of tillage practices on soil microbiome and agricultural parameters," *Science of the Total Environment* 705 (February 25, 2020): 135791.

107 **only beginning to understand:** Stefan Geisen, "The Future of (Soil) Microbiome Studies: Current Limitations, Integration, and Perspectives," *mSystems 6,* no. 4 (August 24, 2021): e00613–21.

107 **soil microbiome and human microbiome:** Heribert Hirt, "Healthy soils for healthy plants for healthy humans," *EMBO Reports* 28, no. 8 (August 5, 2020): e51069.

107 **even our lungs:** Agnieszka Magryś, "Microbiota: A Missing Link in the Pathogenesis of Chronic Lung Inflammatory Diseases," *Polish Journal of Microbiology* 70, no. 1 (March 2021): 25–32.

107 **of autoimmune diseases:** F. De Luca and Y. Shoenfeld, "The microbiome in autoimmune diseases," *Clinical & Experimental Immunology* 195, no. 1 (January 2019): 74–85.

107 **glyphosate:** Michelle Perro and Vincanne Adams, *What's Making Our Children Sick: How Industrial Food Is Causing an Epidemic of Chronic Illness, and What Parents (and Doctors) Can Do about It* (White River Junction, VT: Chelsea Green Publishing, 2017).

107 **2,000 percent increase:** Christopher Walljasper and Ramiro Ferrando, "Use of Controversial Weed Killer Glyphosate Skyrockets on Midwest Fields," *Illinois Public Media*, May 28, 2019.

107 **nutritional loss:** Roddy Scheer and Doug Moss, "Dirt Poor: Have Fruits and Vegetables Become Less Nutritious?" *Scientific American*, April 27, 2011.

108 **five principles of soil health:** Jay Fuhrer, *Soil Health*, USDA Natural Resources Conservation Service, September 2022.

108 **plastic mulch:** Jennifer Tucker, "Memorandum to the National Organic Standards Board," USDA Agricultural Marketing Service, September 9, 2023.

Chapter Six. Black Snake Farm

115 **ancestral homeland:** Catawba Indian Nation, "About the Nation," catawba.com.

115 **archaeologists have traced:** Stephen R. Claggett, "North Carolina's First Colonists: 12,000 Years Before Roanoke," Office of State Archaeology, North Carolina State Historic Preservation Office, March 15, 1996.

115 **tens of thousands of years earlier:** Paul Rincon, "Earliest evidence for humans in the Americas," *BBC News*, July 22, 2020.

116 **heterospecific pollen:** Gerardo Arceo-Gómez, Amelia Schroeder, Cristopher Albor, et al, "Global geographic patterns of heterospecific pollen receipt help uncover potential ecological and evolutionary impacts across plant communities worldwide," *Scientific Reports* 9, no. 8086 (2019).

116 **98 percent:** Jess Gilbert, Spencer D. Wood, and Gwen Sharp, "Who Owns the Land? Agricultural Land Ownership by Race/Ethnicity," *Rural America* 17, no. 4 (Winter 2002): 55–62.

116 **54 percent:** US Department of Agriculture, "Farm Labor," USDA Economic Research Service, Aug. 7, 2023.

117 **over hunting grounds in ancient times:** Allison Entrekin, "Casino Wars in the Carolinas: Two tribes that have battled for centuries are at odds once again," *Atlanta*, April 6, 2020.

117 **intermarried:** Matthew T. Gregg and Melinda C. Miller, "Race and agriculture during the assimilation era: Evidence from the Eastern Band of Cherokee Indians," *Demographic Research* 46, no. 37 (June 21, 2022): 1109–1136.

118 **Mormon missionaries:** "Catawba," Encyclopedia.com, May 29, 2018 (accessed May 20, 2023).

118 **For the Catawba . . . encompasses 700 acres:** Catawba Indian Nation, "About the Nation"; James H. Merrell, *The Indians' New World: Catawbas and Their Neighbors from European Contact through the Era of Removal* (Chapel Hill: Omohundro Institute and University of North Carolina Press, 2010).

120 **omission of post-1890s indigenous history:** Joshua Ward Jeffery, "Why Do Native People Disappear From Textbooks After the 1890s?" *EducationWeek*, August 16, 2021.

122 **the "ecological Indian":** Dina Gilio-Whitaker, "The Problem with the Ecological Indian Stereotype," *Tending the Wild*, PBS SoCal, February 7, 2017.

122 **subhuman wilderness creatures:** John R. Knott, *Imagining Wild America* (Ann Arbor: The University of Michigan Press, 2002).

123 **Diabetes is more prevalent:** Office of Minority Health, "Diabetes and American Indians/Alaska Natives," US Department of Health and Human Services.

123 **one in three Americans:** CDC Newsroom, "Number of Americans with Diabetes Projected to Double or Triple by 2050," Centers for Disease Control and Prevention, October 22, 2010.

Chapter Seven. FarmaSis

129 **more than a foot of rise:** Chloe Johnson, "A missing piece in Charleston's sea level rise puzzle is what's happening on land," *Post and Courier*, September 21, 2020.

129 **"nuisance" flood days:** John H. Tibbetts, "When a City Stops Arguing About Climate Change and Starts Planning," *Next City*, November, 7, 2016.

130 **half a million new residents:** Samuel Stebbins, "Charleston, SC, Will Be Among the Fastest Growing Cities by 2060," *24/7 Wall St.*, September 16, 2022.

130 **Gullah Geechee:** "The Gullah Geechee People," Gullah Geechee Cultural Heritage Corridor Commission, gullahgeecheecorridor.org.

131 **12 million acres:** Vann R. Newkirk II, "The Great Land Robbery," *Atlantic*, September 2019.

131 **$50 billion a year:** Joseph C. Von Nessen, *The Economic Impact of Agribusiness in South Carolina*, South Carolina Department of Agriculture, November 2022.

131   **90 percent of its food:** Nikki Seibert Kelley, "Cultivating Connections: Building a strong food system from farm to table," *Wit Meets Grit*, October 10, 2022.

132   **less than $19,000:** Dustin Waters, "Chicora-Cherokee residents, Palmetto Railways work to lessen new railyard's impact," *Charleston City Paper*, May 11, 2016.

133   **eleven of these bereft:** Paul Bowers, "The poorest neighborhoods have easy access to junk food, but not fresh fruit and veggies," *Charleston City Paper*, December 14, 2011.

133   **Millicent Brown:** Stephanie Hunt, "Somebody Had to Do It," *Charleston*, February 2016.

133   **Obesity, diabetes, high blood pressure:** Office of Minority Health, "Obesity and African Americans," US Department of Health and Human Services.

135   **food hubs:** Healthy Food Access, "Food Hubs," healthyfoodaccess.org.

137   **extant medical racism:** Kat Stafford, Aaron Morrison, and Annie Ma, "Medical Racism in History," *AP News*, May 23, 2023.

138   **Four out of five:** Office of Minority Health, "Obesity and African Americans."

138   **thirteen (proven) types of cancer:** Division of Cancer Prevention and Control, "Obesity and Cancer," Centers for Disease Control and Prevention (reviewed August 9, 2023).

138   **highest rate of maternal mortality:** Donna L. Hoyert, Divison of Vital Statistics, "Maternal Mortality Rates in the United States, 2021," CDC National Center for Health Statistics (reviewed March 16, 2023).

138   **COVID-19 pandemic:** Tamara Rushovich, Marion Boulicault, Jarvis T. Chen, et al, "Sex Disparities in COVID-19 Mortality Vary Across US Racial Groups," *Journal of General Internal Medicine* 36 (April 5, 2021): 1696–1701.

138   **at a time when less than 1 percent:** Elizabeth Becker, "Caucus Seeks More Aid for Black Farmers," *New York Times*, July 11, 2002.

138   **1.2 percent of farmers:** US Department of Agriculture, *2022 Census of Agriculture*, USDA National Agricultural Statistics Service, February 2024.

138   **0.5 percent of organic growers:** Alice K. Formiga, "Statistics on Ethnicity and Race on Organic Farms in the United States," *eOrganic*, August 9, 2021.

138 **only 0.5 percent of farmland:** US Department of Agriculture, *2021 Organic Survey*, USDA National Agricultural Statistics Service, December 2022.

138 **14 percent of American farmers:** Vera J. Banks, *Black Farmers and Their Farms* (Rural Development Research Report Number 59), USDA Economic Research Service, July 1986.

138 **Organic Certification Cost Share Program:** "USDA Restores Organic Certification Cost Share Program," National Organic Coalition, May 10, 2023.

138 **$16 billion:** Tara O'Neill Hayes and Katerina Kerska, "Primer: Agricultural Subsidies and Their Influence on the Composition of US Food Supply and Consumption," American Action Forum, November 3, 2021.

139 **25 percent of this 0.5 percent:** Emma Layman and Nicole Civita, "Decolonizing agriculture in the United States: Centering the knowledges of women and people of color to support relational farming practices," *Agriculture and Human Values* 39 (January 27, 2022): 965–978.

140 **$1 billion endowment:** Stephanie Mirah, "Clemson University endowment surpasses $1B as officials set sights on matching ACC peers," *Post & Courier*, October 5, 2021.

143 **beginning farmer:** US Department of Agriculture, "Limited Resource Farmer and Rancher (LRF/R) – Beginning Farmer Program Definition," USDA Natural Resources Conservation Service.

144 **Project Drawdown:** Project Drawdown, "Regenerative Annual Cropping," drawdown.org.

Chapter Eight. Grass Nomads

150 **in the midst of a megadrought:** Chelsea Harvey and E&E News, "Western 'Megadrought' Is the Worst in 1,200 Years," *Scientific American*, February 15, 2022.

151 **the largest wildfires in New Mexico history:** Amanda Holpuch, "US Forest Service Planned Burn Caused Largest New Mexico Wildfire," *New York Times*, May 28, 2022.

153 **agroecology:** FAO, *The 10 Elements of Agroecology: Guiding the Transition to Sustainable Food and Agricultural Systems*, Food and Agriculture Organization of the United Nations, 2018.

153 **Unschooling:** Laura Monteverdi, "It's legal in all 50 states, but what exactly is unschooling?" THV11, August 9, 2018.

153　**Forty-nine percent:** "Farming in the US," *American Experience*, PBS, PBS.org.

153　**compulsory public education:** Douglas J. Slawson, *The Department of Education Battle, 1918–1932: Public Schools, Catholic Schools, and the Social Order* (Notre Dame: University of Notre Dame Press, 2005).

154　**millions-year-old landscapes:** Jeremy M. B. Smith, "Ecological region: Savanna," *Encyclopedia Britannica*, britannica.com (accessed June 27, 2023).

154　**a pointed tack of genocide:** J. Weston Phippen, "'Kill Every Buffalo You Can! Every Buffalo Dead Is an Indian Gone,'" *Atlantic*, May 13, 2016.

154　**garnered controversy:** David D. Briske, Brandon T. Bestelmeyer, Joel R. Brown, et al, "The Savory Method Can Not Green Deserts or Reverse Climate Change," *Rangelands* 35, no. 5 (October 2013): 72–74.

155　**George Monbiot:** Stefan Gates and George Monbiot, "George Monbiot: Protein production must move from farm to factory," May 27, 2022, in *Food Matters Live*, podcast.

155　**25 percent of all land:** Gregory P. Asner, Andrew Elmore, Lydia Olander, et al, "Grazing systems, ecosystem responses, and global change," *Annual Review of Environment and Resources* 29, no. 26 (November 2004): 261–299.

155　**has degraded 20 to 35 percent:** Liming Lai and Sandeep Kumar, "A global meta-analysis of livestock grazing impacts on soil properties," *PLoS One* 15, no. 8 (August 7, 2020): e0236638.

155　**managed grazing:** Project Drawdown, "Managed Grazing," drawdown. org.

155　**Concentrated animal feeding operations (CAFOs):** Carrie Hribar and Mark Schultz, *Understanding Concentrated Animal Feeding Operations and Their Impact on Communities*, National Association of Local Boards of Health, 2010.

157　**average age of a US rancher:** US Department of Agriculture, *2022 Census of Agriculture*, USDA National Agricultural Statistics Service, February 2024.

157　**95 percent of land:** Texas Land Conservancy, "Why Conserve Land?" texaslandconservancy.org.

157　**155 million rural acres:** Bureau of Land Management, "Livestock Grazing on Public Lands," US Department of the Interior.

158 **damaged by overgrazing:** Public Employees for Environmental Responsibility, "America's Rangelands Deeply Damaged by Overgrazing," March 5, 2020.

Chapter Nine. Rooted Northwest

163 **more than 200 agrihoods:** Matthew Norris, *Agrihoods: Cultivating Best Practices,* Urban Land Institute, 2018.

164 **guided by *biophilia*:** Edward O. Wilson, *Biophilia: The Human Bond with Other Species* (Cambridge, MA: Harvard University Press, 1984).

164 **death rates from suicide, alcohol, drug overdoses:** Erika Edwards, "US death rates from suicide, alcohol, and drug overdoses reach all-time high," NBC News, June 11, 2019.

164 **Seventy-four percent:** Cheryl D. Fryar, Margaret D. Carroll, and Joseph Afful, "Prevalence of overweight, obesity, and severe obesity among adults aged 20 and over: United States, 1960–1962 through 2017–2018," CDC National Center for Health Statistics (Health E-Stats), 2020.

164 **20 percent of children and teens:** CDC, "Childhood Obesity Facts (2017–2020)," National Center for Chronic Disease Prevention and Health Promotion (reviewed May 17, 2022).

164 **a hundred pounds of sugar:** Susan Ratz, "The Question of Sugar," USDA Agricultural Research Service, September 11, 2023.

164 **1,500 miles:** Becky Henne, "How far did your food travel to get to you?" Michigan State University Extension, September 20, 2012.

165 **$1,000 monthly stipend:** Taylor Reid, "Organic Farm Apprenticeship in Georgia," *Beginning Farmers,* July 1, 2021.

166 **175 acres every hour:** GAP Initiative, "America's Disappearing Farm and Range Land," GAP Initiative at Virginia Tech.

166 **70 percent of land is conserved:** Code of Ordinances City of Chattahoochee Hills, GA, "Article XIII Sec. 13-3: TDR, PDR, and DTC programs purpose," version May 10, 2023.

167 **One in four US golf courses:** Mary Richardson, "Golf course living is appeal of these communities," *Gainesville Sun,* October 13, 1985.

167 **local food systems we could create:** AJ Dellinger, "Agrihoods imagine a future organized around community agriculture," *Mic,* April 22, 2022.

167 **FEMA Special Flood Hazard:** "FEMA Flood Map Service Center," Federal Emergency Management Agency; "FEMA Flood Maps and

Zones Explained," Federal Emergency Management Agency, April 4, 2018 (updated July 28, 2023).

169 **cohousing:** Shane Madden, "The History of Co-Operative Housing in Denmark and the impact of Cultural Changes over time," OIKONET, www.oikodomos.org.

172 **56 percent of American farmers:** Jesse Hirsch, "Most Farmers Need Second Jobs to Survive," *Ambrook Research*, September 29, 2022.

173 **successful (not all regenerative) farming co-ops:** US Department of Agriculture, "Top 100 Largest Agricultural Cooperatives – rank, name, type, total business volume and total assets, 2014 and 2013," USDA Rural Development.

Chapter Ten. Jewish Farmer Network

178 **first federal definition of bullying:** Paul Timm, *School Security: How to Build and Strengthen a School Safety Program* (Butterworth-Heinemann, 2021).

180 **criticism for labor issues:** Martina Žoldoš, "My WWOOFing Experience Was a Nightmare. Here's Why the Organization Needs to Change," *Matador Network*, October 17, 2016.

182 **earthworks:** Douglas Barnes, *The Permaculture Earthworks Handbook: How to Design and Build Swales, Dams, Ponds, and Other Water Harvesting Systems* (New Society Publishers, 2017).

183 **Saguaro and prickly pear fruit:** Chris Malloy, "The Ancient Art of Harvesting Fruit in the Desert," Gastro Obscura, *Atlas Obscura*, September 25, 2020.

183 **Tohono O'odham people:** "Native Peoples of the Sonoran Desert: The O'odham," National Park Service (updated January 23, 2021).

184 **"cooperative overlapping":** Deborah Tannen, *Conversational Style: Analyzing Talk among Friends* (London: Oxford University Press, 2005).

184 **Pale of Settlement:** "Modern Jewish History: The Pale of Settlement," Jewish Virtual Library (accessed July 1, 2023).

184 **"Jews run Hollywood" echoism:** Andrew Lapin, "Dave Chappelle isn't the first to suggest that Jews run Hollywood. Here are the origins of the trope," *Jewish Telegraphic Agency*, November 16, 2022.

185 **a renewable agriculture even in antiquity:** David R. Montgomery,

Dirt: The Erosion of Civilizations (Oakland: University of California Press, 2012).

186 **for inner-city Jewish immigrant kids:** Jewish Federation of Detroit, "Scenes from the Archives," *My Jewish Detroit*, August 1, 2012.

189 **herding in Central Asia:** William T. T. Taylor, Mélanie Pruvost, Cosimo Posth, et al, "Evidence for early dispersal of domestic sheep into Central Asia," *Nature Human Behavior* 5, no. 9 (2021), 1169–1179.

Additional Sources

Fukuoka, Masanobu, *The One-Straw Revolution: An Introduction to Natural Farming* (New York: New York Review of Books, 1978).

Hayes, Shannon, *Radical Homemakers: Reclaiming Domesticity from a Consumer Culture* (Walnut Creek, CA: Left to Write Press, 2010).

Hawken, Paul, ed., *Drawdown: The Most Comprehensive Plan Ever Proposed to Reverse Global Warming* (New York: Penguin Books, 2017).

Hawken, Paul, *Regeneration: Ending the Climate Crisis in One Generation* (New York: Penguin Books, 2021).

Kingsolver, Barbara, *Animal, Vegetable, Miracle: A Year of Food Life* (New York: HarperCollins Publishers, 2007).

Little, Amanda, *The Fate of Food: What We'll Eat in a Bigger, Hotter, Smarter World* (New York: Harmony Books, 2019).

Miller, Daphne, *Farmacology: Total Health from the Ground Up* (New York: William Morrow, 2013).

Mock, Sarah, *Farm (And Other F Words): The Rise and Fall of the Small Family Farm* (District of Columbia: New Degree Press, 2021).

Montgomery, David R., *Dirt: The Erosion of Civilizations* (Berkeley and Los Angeles: University of California Press, 2008).

Penniman, Leah, *Farming While Black: Soul Fire Farm's Practical Guide to Liberation on the Land* (White River Junction, VT: Chelsea Green Publishing, 2018).

Philpott, Tom, *Perilous Bounty: The Looming Collapse of American Farming and How We Can Prevent It* (New York: Bloomsbury Publishing, 2020).

Smaje, Chris, *A Small Farm Future: Making the Case for a Society Built Around Local Economies, Self-Provisioning Agricultural Diversity, and a Shared Earth* (White River Junction, VT: Chelsea Green Publishing, 2020).

Sundeen, Mark, *The Unsettlers: In Search of the Good Life in Today's America* (New York: Riverhead Books, 2017).

Thapar, Neil, "An Enormous Land Transition is Underway. Here's How to Make it Just," Civil Eats, February 24, 2020, http://tinyurl.com/cxfp2ta3.

Tickell, Josh, *Kiss the Ground: How the Food You Eat Can Reverse Climate Change, Heal Your Body & Ultimately Save Our World* (New York: Atria/ Enliven Books, 2017).

Tree, Isabella, *Wilding: The Return of Nature to a British Farm* (London: Picador, 2018).

INDEX